AF784539

HANDBOOK OF INDUSTRIAL FOOD MICROBIOLOGY

HANDBOOK OF INDUSTRIAL FOOD MICROBIOLOGY

Manju Nehra, PhD
Vikash Nain, PhD

APPLE ACADEMIC PRESS

First edition published 2025

Apple Academic Press Inc.
1265 Goldenrod Circle, NE,
Palm Bay, FL 32905 USA

760 Laurentian Drive, Unit 19,
Burlington, ON L7N 0A4, CANADA

CRC Press
2385 NW Executive Center Drive,
Suite 320, Boca Raton FL 33431

4 Park Square, Milton Park,
Abingdon, Oxon, OX14 4RN UK

© 2025 by Apple Academic Press, Inc.

Apple Academic Press exclusively co-publishes with CRC Press, an imprint of Taylor & Francis Group, LLC

Reasonable efforts have been made to publish reliable data and information, but the authors, editors, and publisher cannot assume responsibility for the validity of all materials or the consequences of their use. The authors are solely responsible for all the chapter content, figures, tables, data etc. provided by them. The authors, editors, and publishers have attempted to trace the copyright holders of all material reproduced in this publication and apologize to copyright holders if permission to publish in this form has not been obtained. If any copyright material has not been acknowledged, please write and let us know so we may rectify in any future reprint.

Except as permitted under U.S. Copyright Law, no part of this book may be reprinted, reproduced, transmitted, or utilized in any form by any electronic, mechanical, or other means, now known or hereafter invented, including photocopying, microfilming, and recording, or in any information storage or retrieval system, without written permission from the publishers.

For permission to photocopy or use material electronically from this work, access www.copyright.com or contact the Copyright Clearance Center, Inc. (CCC), 222 Rosewood Drive, Danvers, MA 01923, 978-750-8400. For works that are not available on CCC please contact mpkbookspermissions@tandf.co.uk

Trademark notice: Product or corporate names may be trademarks or registered trademarks and are used only for identification and explanation without intent to infringe.

Library and Archives Canada Cataloguing in Publication

Title: Handbook of industrial food microbiology / Manju Nehra, PhD, Vikash Nain, PhD.
Other titles: Industrial food microbiology
Names: Nehra, Manju, author. | Nain, Vikash, author.
Description: First edition. | Includes bibliographical references and index.
Identifiers: Canadiana (print) 20240343174 | Canadiana (ebook) 20240343204 | ISBN 9781774915264 (hardcover) | ISBN 9781774915271 (softcover) | ISBN 9781003412779 (ebook)
Subjects: LCSH: Food—Microbiology—Handbooks, manuals, etc. | LCGFT: Handbooks and manuals.
Classification: LCC QR115 .N44 2025 | DDC 664.001/579—dc23

Library of Congress Cataloging-in-Publication Data

CIP data on file with US Library of Congress

ISBN: 978-1-77491-526-4 (hbk)
ISBN: 978-1-77491-527-1 (pbk)
ISBN: 978-1-00341-277-9 (ebk)

About the Authors

Manju Nehra, PhD
Chairperson, Department of Food Science and Technology, Chaudhary Devilal University, Sirsa, Haryana, India

Manju Nehra, PhD, is the chairperson of the Department of Food Science and Technology at Chaudhary Devilal University, Sirsa (Haryana, India). She has been an academician throughout her professional career, which spans around 14 years. Since 2007, she has been teaching as a faculty member in the Department of Food Science and Technology at Chaudhary Devilal University, Sirsa. She is actively working on new product development from traditional Indian crops, fruits, and vegetables and improving the livelihood of farmers and rural ladies by providing knowledge about minimal processing and preserving techniques through lectures and workshops. She has participated in various national and international conferences and presented her research work in oral talks and posters. She has more than 40 research papers in journals of national and international repute. She actively contributes to university administration by providing services as Director of Youth Welfare. She has five books to her credit. She has a personal website to connect with the public to learn more about the changing scenario of human nutrition.

She is associated with various professional bodies like AFSTI, AMI, and Punjab Science Congress and provides consultancy services to various agro-based industries. She is actively involved in curriculum development and nutrition counseling. She is profoundly interested in teaching, student counseling, environment conservation, and traveling, besides hygiene and nutrition.

She completed her BSc in the medical stream from Kurukshetra University, Kurukshetra, and her MSc in Food Technology from Guru Jambheshwar University of Science and Technology, Hisar. Her doctorate was also from GJUS$T, Hisar. Her research work was related to the efficacy of natural antioxidants in enhancing the shelf life of oils.

Vikash Nain, PhD
Assistant Professor (contract), Food Science and Technology Department
Chaudhary Devilal University, Sirsa, Haryana, India

Vikash Nain, PhD, is working as an Assistant Professor (contract) in the Food Science and Technology Department at Chaudhary Devilal University. He has also worked as a teaching associate in the Department of Food Technology at Guru Jambheshwar University of Science and Technology, Hisar. Dr. Nain has various national and international publications to his credit. He has contributed valuable suggestions and guidance in mentoring post-graduate students' research projects to fulfill their academic requirements. Presently, Dr. Nain is working on developing and characterizing starch nanoparticles and their application in biodegradable films, value addition in product development such as protein-rich new product development, and incorporating indigenous medicinal components in traditional foods. He has more than 10 years of experience in the field of Food Science and Technology and serves as a consultant for the local food industries, with an emphasis on food sanitation and food safety.

Contents

Abbreviations

ALP	alkaline phosphatase
BCR	bubble column reactor
BSE	bovine spongiform encephalopathy
CH4	methane
CLEA	cross-linking enzyme aggregate
CLEC	cross-linking enzyme crystal
CMC	carboxymethyl cellulose
CO2	carbon dioxide
DEAE	diethylaminoethyl
DEAE-C	diethylaminoethyl cellulose
EPA	Environmental Protection Agency
ETC	electron transport chain
FAO	Food and Agriculture Organization
FDA	Food and Drug Administration
H2	hydrogen
HFCS	high fructose corn syrup
HUS	hemolytic uremic syndrome
IBS	irritable bowel syndrome
LA	lactic acid
LAB	lactic acid bacteria
MBR	membrane bioreactors
MFGM	milk fat globule membrane
N2	nitrogen
NMWCO	nominal molecular weight cut-off
NRRL	Northern Regional Research Laboratories
PAN	poly acrylo nitrile
PTFE	polytetrafluoroethylene
PVA	polyvinyl alcohol
RCF	relative centrifugal force
SCP	single-cell protein
SDS	sodium dodecyl sulfate
SSF	solid-state fermentation
TD	thermal destruction
WHO	World Health Organization

Acknowledgment

When the entire atmosphere around us encourages learning, learning is inevitable. We enormously enjoyed writing this book. It gave us a chance to explore our abilities as well as produce a valuable work. We write here today thanking the Almighty for his immense blessings at all times, and which enabled us to do even better in this work. The most powerful potential of a human being manifests when he finds strength in the cosmic topsoil where there is infinite to understand and comprehend. With every project comes enlightenment; we have been enlightened that nothing lasts forever and that there is an ordinary way of doing different tasks.

We are indebted to our learnings from the present and past, our parents, siblings, and mentors. We appreciate the time they invested in helping us pull this project. Whole-hearted gratitude to Mr. Vikram Nehra, Mrs. Sunita Sangwan, and Dr. Aruna Sharma, who have been our pillars of strength and always stood by our side. A special thanks to friends who played a significant role in completing the project through their constant encouragement.

We thank our entire families who prayed for our wellness during this time and motivated us to stay determined and focussed. During certain some tough times, we may find that our light goes out. However, it is rekindled with a spark from another person; each of us has reason to think with deep gratitude of those who have lighted the flame within us.

We are also thankful to the authorities of Chaudhary Devi Lal University, Sirsa, India, for their constant motivation and encouragement during the entire service period. Last but not the least, the authors are grateful to the publisher and editorial members for accepting the proposal and editing the chapters.

We thank all the people who have contributed to completing this venture in one way or another.

—***Editors***

Preface

Compared to older biological sciences like zoology and botany, where of bacteria were first discovered more than 300 years ago, the field of microbiology is yet in a continually developing phase. Microbiology deals with the science of microbes, which are also referred to as single-celled or cell-cluster microflora. These are largely differentiated into two groups, i.e., prokaryotes and eukaryotes. In brief, the study of life of microscopic organisms that are invisible to the unaided eye is known as microbiology.

Industrial microbiology is largely concerned with the commercial use of microbes and involves processes and products that are extremely significant globally in terms of economics, the environment, and society. Industrial microbiology has two main facets, the first of which is the fermentation-based synthesis of useful microbial compounds. These include age-old fermented foods and drinks, including bread, beer, cheese, and wine, that have been made for millennia. Microorganisms have also been used in the past century or so to produce a wide range of chemical feedstocks, energy sources, enzymes, food additives, and medications. The second is how microorganisms are used to provide services, especially for waste treatment and pollution management, which makes use of their capacity to destroy almost all natural materials. However, such activities must be controlled while these materials are in use. Otherwise, the consequent bio-deterioration will lead to major economic losses.

The goal of this textbook is to introduce industrial microbiology. The authors' experience instructing industrial microbiology and other areas of applied microbiology served as a foundation for their writing. This book is intended for undergraduates and beginning graduate students in chemical engineering, microbiology, and biotechnology. It will be interesting to those who are studying pharmacy, biochemistry, and general biology. Students and professionals in the fields of civil engineering and public health will be interested in the section on waste disposal. Elements of biology and microbiology are introduced for the benefit of students who may need to become more familiar with the basic concepts of biological sciences that act as the base of industrial microbiology, for example, graduates of chemical and civil engineering. The various applications of industrial microbiology are broadly covered, and the chapters are grouped to make it easier to follow.

Introduction

Chapter 1 of this book deals with the introduction to industrial microbiology and its areas of application. This chapter gives insights into the introductory aspects for the new readers of microbiology. Chapter 2 describes the historical breakthrough in basic and industrial microbiology that changed the prospects of its applications in diversified fields. In this chapter, both ancient and modern major developments are described. Chapter 3 deals with the microbiology of industrial microorganisms and the public health aspects of microbes. Food poisoning, indicator microorganisms and factors effecting the growth of microorganisms are discussed for basic knowledge of microbial growth and safety purposes. Chapter 4 deals with the most important unit operation used in industrial bio-production, i.e., bioreactors. Here, industrial fermenters, their design, types, and operating conditions are discussed in detail. It is important to understand the operating design and conditions before going for industrial production of commercially important products.

Chapter 5 describes the raw material required in the fermentation system. Here, ideal characteristics and useful raw materials for the industrial setup of such a fermentation system are discussed in detail. The suitability of raw materials for different fermentation systems is also detailed. Various primary metabolites of microbial origin are being commercially produced and widely used in food and feed manufacturing. These consist of amino acids, alcohols, organic acids, vitamins, flavor modulator nucleotides, polyols, polysaccharides, and sugars. Their production is discussed in detail in Chapter 6 of this book.

After producing the product, extraction, and purification remain the next most important steps in biological processes. These necessary steps are incredibly time-consuming and costly. Steps involved in the recovery and purification are discussed in Chapter 7 of the book under the title, Downstream Processing. The recovery price is determined by the nature and purity of the product and the recovery methods used. Secondary metabolites from microbial sources significantly impact our society, including human health, nutrition, and economics. These include immune-modulating agents, antibiotics, enzyme inhibitors, antagonists and agonists receptors, pigments, antitumor agents, pheromones, growth regulators of animals and plants, and many other compounds. Their production and methods for over productions are discussed in detail in Chapter 8.

Biomass is among the most available renewable sources of energy on planet Earth. Their industrial productions for different applications are discussed in the next chapter under the title, Biomass Production. Enzymes are used as eco-friendly biocatalysts in many industries, such as chemicals, pharmaceuticals, and food industries, for commercial production and as biosensors for target analysts' identification or determination. Chapter 10 deals with the immobilized enzyme, their microbial production, and their application in various commercial activities. The last chapter of this book deals with the application of microbes in wastewater treatments. Waste is an inevitable by-product of the majority of human action. Economic growth and increasing living standards have resulted in a rise in the quantity and complexity of waste generated. This can be treated as commercially important by-products with the help of microbes. Their application in this field is described in detail in this chapter.

CHAPTER 1

Introduction to Industrial Microbiology

Industrial microbiology is a field that has previously been recognized. It has been acknowledged in the present time as well. Transformations have occurred in this field. The modifications have been primarily brought about by utilizing new tools and technologies. This chapter covers the biological control of insects, bacteria, and fungus that affect different crops. Many microorganisms, such as bacteria, algae, viruses, and fungi, can biologically control these severe issues. The manner of action of the various microorganisms employed to control insects and plant diseases is described in detail in this chapter. Industrial microbiology has traditionally been recognized as a distinct discipline. It has also been acknowledged in recent years. In this field, there have also been changes. The reforms were mainly achieved through the application of new techniques and technology. The widespread use of bioprocesses has presented an opportunity for many engineers to advance in biotechnology. The utilization of microbes to manufacture alcohols and acetone, which are employed in industrial processes, is among the most useful applications of biotechnology.

Industrial microbiology, also known as microbial biotechnology, implements engineering principles to the processing and packaging by microorganisms like bacteria, fungi, algae, protozoa, viruses, and plant and animal cells to produce valuable products or processes. The micro-organisms utilized may be native isolates, laboratory-selected mutants, or microbes that have been genetically modified using recombinant DNA methods (Okafor et al., 2007).

The practice of industrial microbiology or biotechnology is a very old one. Long before their discovery, microorganisms were exploited to benefit human beings.

The use of microorganisms in producing many products, including enzymes, foods, beverages, chemical bioproducts, fuels, pharmaceuticals, and clean technologies for waste treatment and pollution control, is of major economic, environmental, and social importance. Microorganisms are used

Handbook of Industrial Food Microbiology. Manju Nehra & Vikash Nain
© 2025 Apple Academic Press, Inc. Co-published with CRC Press (Taylor & Francis)

to produce essential substances such as antibiotics, food products, enzymes, amino acids, vaccines, and active pharmaceutical ingredients in industrial microbiology, a branch of applied microbiology (Shadia et al., 2017).

Industrial microbiology has served humanity since prebiblical times, providing fermented beverages and foods to enhance the quality of life. Thanks to advances in industrial microbiology and chemical purification, the isolation of enzymes in commercial quantities has been made possible. They are currently widely used in a variety of sectors, including winemaking (Godfrey et al., 1966).

In industrial microbiology, microbial metabolites can be employed to obtain amino acids, generate vaccines and medicines, and isolate compounds needed for organic synthesis. There are two types of metabolites: main and secondary metabolites. Microbes produce primary metabolites such as carbohydrates, amino acids, proteins, vitamins, ethanol, acetone, organic acids, and others during energy metabolism and are considered essential for microbial growth (Abbott et al., 1979).

In revolutionary exploitation and establishment, industrial microbiology has played a crucial role. Recombinant protein has been produced in a variety of microbes, for example, *Escherichia coli*, *Bacillus subtilis*, *Saccharomyces cerevisiae*, *Pichia pastoris*, *Hansenula polymorpha*, and *Aspergillus niger* (Vieira Gomes et al., 2018).

Industrial microbiology has been practiced since the dawn of time. Microorganisms were used to meet the needs and requirements of humans even before they were discovered, for example, to preserve dairy products like cheese, milk, fruits, and vegetables, and improve the quality of life by making beverages, cheeses, bread, pickled foods, and vinegar. Yeasts have been used since prehistoric times. As early as 7000 B.C., the oldest fermentation to understand the conversion process of sugar to alcohol by yeasts was discovered to make beer in Sumeria and Babylonia. By 4000 B.C., historians claim that the Egyptians had discovered that brewer's yeast could leaven bread by producing carbon dioxide (CO_2) (Najafpour et al., 2007). Molds and germs have also been used by ancient people to make cheese. Another outmoded fermentation product, wine, was invented in Assyria about 3500 B.C., and there is a mention to wine in the book of Genesis, where Noah drank too much of it (Demain et al., 2010).

In contrast to many other aspects of the study of microbiology, the microbiologist in an industrial setting does not work alone. He is usually just one of many different bureaucrats with whom he interacts regularly (Okafor et al., 2007). Some function of microbiologist:

- The organism to be used in the processes is chosen.

- The choice of the growth medium of the organism.
- Determining the environmental conditions for the organism's optimum productivity, i.e., pH, temperature, aeration, etc.
- During manufacturing, the microbiologist must keep an eye on the process for pollutants and assist in quality control to verify that the products are consistent.
- The proper custody of the organisms in a culture collection ensures their desirable properties are retained.
- The enhancement of microorganism performance through genetic manipulation or media reconstitution.

1.1 MICROORGANISMS USED IN INDUSTRIES

Micro-organisms were supposed to be used in industrial processes even before they were known. Since the dawn of time, traditional methods like fermenting beverages and vinegar and leavening bread have been passed down to us. The discovery of micro-organisms with their diversity of particular biochemical activities has stimulated a steady growth of industrial fermentation processes. Some microorganism and their uses are listed in Table 1.1.

TABLE 1.1 List of Microorganisms and Their Uses

Microorganism	Uses
Saccharomyces cerevisiae	Bakery, wine, beer, sake
Saccharomyces carlsbergensis	Light beer
Saccharomyces rouxii	Soy sauce
Lactobacillus sanfrancisco	Sourdough French bread (sour bread)
Streptococcus thermophiles	Yogurt
Aspergillus oryzae	Saké
Rhizopus sp.	Tempeh
Kluyveromyces fragilis	Ethanol (from glucose)
Xanthomonas campestris	Polysaccharides
A. niger	Citric acid

1.1.1 FERMENTATION

Fermentation is a complicated series of processes caused by microbes that can occur simultaneously. Fermentation occurs when yeast and other microbes, such as lactobacilli, work together to produce a product. Collado-Fernández et al. (2003) define fermentation as the action of yeast on simple sugars to produce CO_2, ethanol, and certain aromatic compounds. Different microorganisms produce different fermentation products, determined mainly by the cell's enzyme complex and ambient circumstances. Industrial microbiology arose as a result of the commercial value of these by-products (Doelle et al., 1975).

1.1.2 BAKERY

The taste and aroma are important attributes for the food industry, especially in the bakery and pastry industry. The fermentation and baking processes can influence the flavor of fermented foods in different ways (Balestra et al., 2015). During a lactic acid (LA) fermentation, the most evident change is the production of acid that lowers the pH, thus increasing the sourness (McFeeters et al., 2004). Fermentation in bakery is mainly related to yeast and LA bacteria. Their metabolic activities are responsible for dough quality and hence for much of the quality of final bakery products (Therdthai et al., 2014). Yeast was grown in aseptic systems to ensure only suitable strains could be selected. Today, baker's yeast is regarded as Saccharomyces cerevisiae. CO_2 gas is generated during yeast fermentation (Bounaix et al., 2010).

$$C_6H_{12}O_6 + \text{Yeast} \rightarrow 2CH_3CH_2\,OH + 2CO_2 + \text{yeast}$$

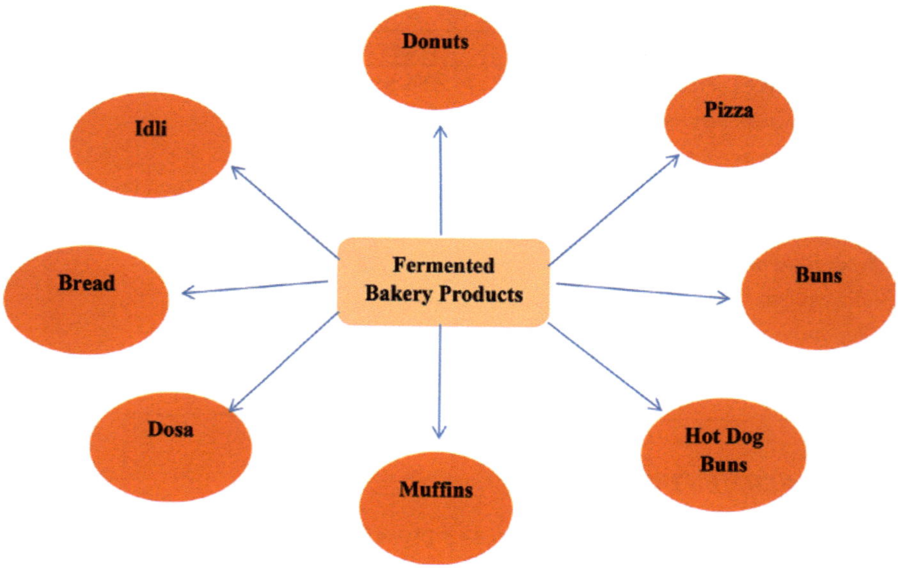

1.1.3 BEVERAGES

For most fermentation processes in the beverage industry, the three yeast species *Saccharomyces cerevisiae*, bayanus, and pastorianus are used; their position within the *S. sensu* stricto complex is explained in the section on the phylogenetics and systematics of yeas (Dequin et al., 2003). Fermentation processes are performed by cells to extract energy from a starting material such as glucose. Alcoholic fermentation is the anaerobic transformation of sugars, mainly glucose and fructose, into ethanol and CO_2 (Constanti et al., 1997) (Table 1.2).

TABLE 1.2 List of Microorganisms Used in Beverage

Microorganisms	Beverage
Acetobacter aceti	Vinegar
Acetobacter cerevisiae	Beer
Acetobacter fabarum	Coffee
Aspergillus acidus	Tea
Cyberlindnera mrakii	Wine
Gluconacetobacter oboediens	Vinegar
Lactobacillus collinoides	Cider
Lactobacillus pentosus	Wine
Saccharomyces bayanus	Beer

Source: Adapted from International Food Biotechnology Council (1990).

1.1.4 FRUIT AND VEGETABLES

Lactic acid (LA) fermentation is predicted to play a significant role in preserving fresh fruits, vegetables, and other food products for feeding people in developing countries (Demir et al., 2006). Fermenting vegetables and fruits with LA are a general procedure for preserving and improving food products' nutritional and sensory qualities. Various traditional naturally fermented meals yielded many viable lactic acid bacteria (LAB) (Anandharaj et al., 2013). *Lactobacillus plantarum*, *Lactobacillus brevis*, *Lactobacillus fermentum*, *Lactobacillus casei*, *Leuconostoc mesenteroides*, *L. kimchi*, *L. fallax*, *Weissella confusa*, *W. koreenis*, *W. cibaria*, and *Pediococcus pentosaceus*, which are considered probiotic sources of the food practice, ferment traditional Asian ferment The availability of certain nutrients such as vitamins and minerals, as well as the acidic nature of fruits and vegetables, creates an ideal environment for LAB fermentation (Heller et al., 2001).

The important fermented fruits and vegetables are categorized as follows:

- Root vegetables such as carrots, turnips, beetroot, radishes, celeriac, and sweet potato (Panda et al., 2007).
- Vegetable fruits including cucumbers, olives, tomatoes, peppers, okra, and green peas (Dahal et al., 2005).
- Vegetables juices such as carrot, turnips, tomato pulp, onion, sweet potato, beet, and horseradish (Panda et al., 2007).
- Fruits like apples, pears, immature mangoes, immature palms, lemons, and fruit pulps such as banana.

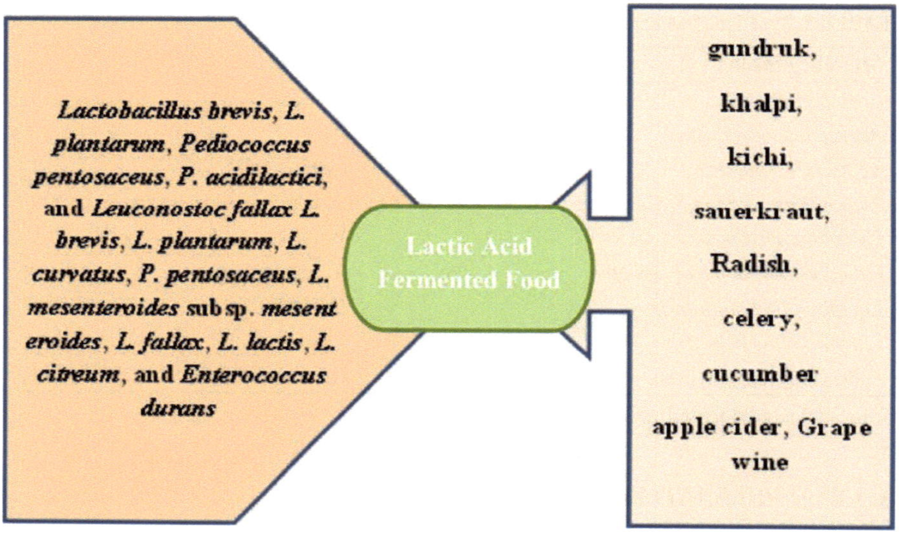

1.1.5 DAIRY INDUSTRY

Dairy products have always been a staple of the human diet. Calcium, vitamin D, proteins, and other vital nutrients are abundant in these foods. Phosphorus, potassium, magnesium, and various vitamins, including vitamin A (retinol), vitamin B12 (cyanocobalamin), and riboflavin, are also present in these foods. Different microbial strains make various fermented dairy products (Kumar et al., 2017). Microbes convert lactose and other carbohydrates in milk to LA and other compounds. Because the acid precipitates the proteins in the milk, fermented products are often thicker than milk. Different types of fermented milk and associated products have been evolved throughout the world, each with its history. The type of milk used, the pre-treatment of the milk, the temperature (climate), the conditions of fermentation, and the subsequent technological treatments all impact their character. Curd, yogurt, cheese, kefir, and kumis are the most regularly utilized dairy products (Pintado et al., 1996).

1.1.6 CURD

Curd is prepared by coagulating or curdling milk. This can be accomplished by incorporating palatable acidic ingredients such as lemon juice or vinegar into the milk. When these ingredients are added to milk, it curdles and separates into two portions. The whey is the liquid portion, and the curd is the solid milk. The whey contains milk whey proteins, whereas the curds contain milk proteins (casein) (De Los Reyes-Gavilán et al., 2015).

1.1.7 YOGURT

It is the most widely consumed dairy product. It's made by boiling milk to about 80°C to destroy any microbes and disintegrate milk proteins. After that, the milk is left to cool down slowly to roughly 45°C before being injected with bacteria and fermented at room temperature. *Lactobacillus delbrueckii* subsp. *bulgaricus* and *Streptococcus salivarius* subsp. Thermophiles were employed as bacteria (Soukoulis et al., 2007). A considerable quantity of yogurt is produced in industry, which is influenced by various elements such as milk quality, milk standardization, milk additions, de-aeration, homogenization, heat treatment, culture selection, and plant design. The milk used to make yogurt has to be of the greatest microbiological purity possible (Lourens-Hattingh et al., 2001).

1.1.8 CHEESE

It is a fermented milk product that has traditionally been used to preserve milk. The coagulation of the milk protein casein results in solid curd and liquid whey formation in the first stage of cheese production. Acidification and proteolysis are two complementary strategies for casein coagulation (Button et al., 2012). LAB and other cheese microorganisms' capacity to break down amino acids into fragrance compounds varies dramatically depending on the strain. These have enzyme systems that allow them to use amino acids in their metabolism (Yvon et al., 2001).

1.1.9 KEFIR

It is a fermented milk beverage that originated in Eastern Europe thousands of years ago. This mild alcoholic beverage is made by inoculating raw milk with kefir grains, irregularly shaped gelatinous white/yellow grains (Ardö et al., 2006). These Kefir grains include yeasts, LAB, acetic acid bacteria, and mycelial fungus in variable and complicated microbial compositions. *Lactobacillus acidophilus*, *Lactobacillus helveticus*, *Lactobacillus casei*, *Lactobacillus fermentum*, *Lactobacillus kefiri*, *Leuconostoc mesenteroides*, *Lactobacillus brevis*, *Lactobacillus lactis*, *Lactobacillus lactis*, *Lactobacillus lactis*, *Lactobacillus lact* (Witthuhn et al., 2005).

1.1.10 KUMIS (KOUMISS)

Kumis and kefir are almost similar dairy products, but kumis is made from a liquid starting culture rather than solid kefir "grains." *Lactobacillus* bacteria are used with restricted acidification characteristics, such as *L. acidophilus*, *L. plantarum*, *L. casei*, and *L. fermentum*. The pH is reduced to 4.5–3.9 at the end of the process, with a conversion ratio of 50% (Danova et al., 2005) (Figure 1.1).

1.1.11 INDUSTRIAL ENZYME

Kefir is a traditional fermented beverage that dates back thousands of years to Eastern Europe. Inoculating raw milk with kefir grains, irregularly shaped gelatinous white/yellow grains produce this moderate alcoholic beverage (Ardö et al., 2006). Yeasts, LA, acetic acid, and mycelial fungus are all present in these Kefir grains in varying and complex microbial compositions (*Lactobacillus brevis*, *Lactobacillus lactis*, *Lactobacillus delbrueckii*).

In the food industry, biocatalysis has been used to produce raw materials and final products for a long time (Fernandes et al., 2010). However, most uses of biocatalysis have focused on hydrolytic reactions for de-branching, improving the solubility, and clarification. In recent years, proteases have arisen as standard biocatalysts in many industrial processes in different fields, such as pharmaceutical, medicine, detergent manufacturing and food science (Jesús Fernández-Lucas et al., 2017) (Table 1.3).

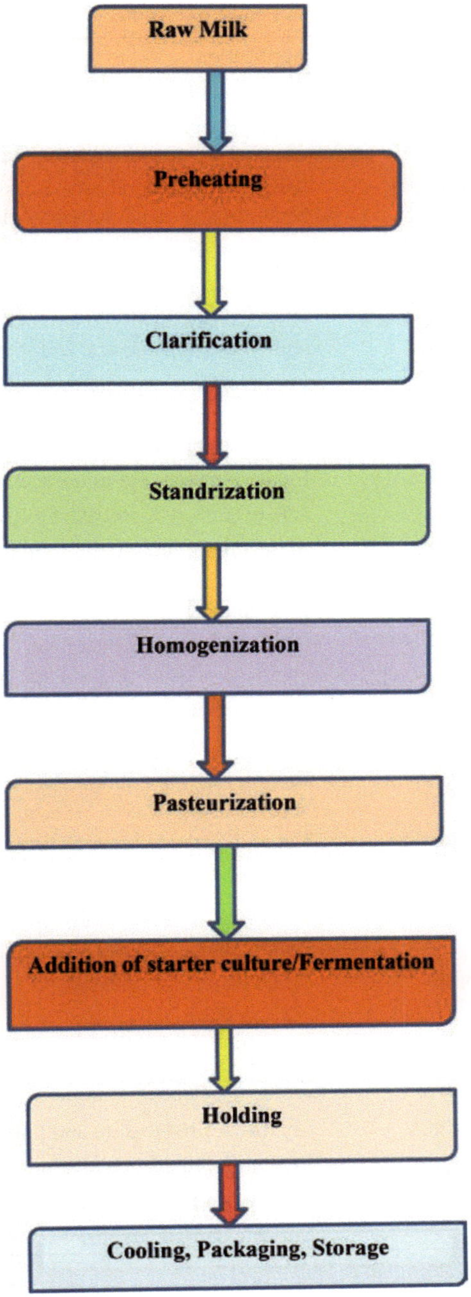

FIGURE 1.1 Flow chart of milk processing.

TABLE 1.3 Some Uses of Enzymes in Food Production

Industry	Enzyme	Application
Baked goods	Lipases	Alteration of the gluten network.
		Higher dough stability, a lower degree of softening and stickiness, and improved crumb structure.
	Amylases	Shorter fermentation period, less viscous dough, better final texture, and greater loaf volume.
		Inhibition of crumb hardening.
		Increase content of slowly digestible starch.
	Hemicellulase	Higher bread quality in terms of dough machinability, crumb structure, and stability.
		Improve sensory properties and volume of bread.
		Remove the insoluble arabinoxylans interfering with the formation of gluten
		Enhance arabinoxyl-oligosaccharides.
		Life extension of prebaked rolls.
		Reduction of mixing time and dough consistency.
	Lipoxygenases	Promote dough raising and foaming and diminish time mixing.
		Bleach carotenoids in doughs.
	Glucose oxidases	Oxidizing agent replacing the carcinogenic potassium bromate.
	Asparaginase	Remove carcinogenic acrylamide.
	Laccases	Biopolymer cross-linker that yields a stronger arabinoxylan network.
Oleochemistry	Lipases	Natural synthesis of emulsifiers.
		Increase properties of low nutritional value oils.
		Produce *trans*-free margarines and shortening.
		Improve fatty acid stability and lipophilicity.
Dairy	Lipases	Speed up cheese ripening.
		Control flavor in cheese.
		Imitate human milk fat to create infant formula.
	Lactases	Remove lactose from milk.
	Proteases	Control desired texture and flavor.
		Ease milk protein coagulation.
		Achieve correct ripening.
		Mitigate dairy allergens.

Source: Adapted from Sanromán & Deive (2017).

1.1.12 ANTIMICROBIAL METABOLITE

In industrial microbiology, microbial metabolites can obtain amino acids, generate vaccines and medicines, and isolate compounds needed for organic synthesis. There are two types of metabolites-Primary and secondary metabolites (Kumar et al., 2004). In organisms' growth, development, and reproduction, primary metabolites are also known as core metabolites. Secondary metabolism is the process that results in the generation of secondary metabolites, and it has been linked to sporulation and colony formation in microbes and has been discovered to begin during the stationary phase of microbial development (Maplestone et al., 1992; Stone & Williams, 1992).

Various metabolites produced in industrial fermentation are discussed in subsections (Prasanna et al., 2021).

1.1.12.1 AGRICULTURE APPLICATION

Due to various fertilizers and insecticides, the need for agricultural products is always expanding. Overuse of chemical fertilizers and pesticides has long-term consequences. The soil becomes infertile and unfit for producing crops due to the overuse of chemical fertilizers and pesticides. Biofertilizers, biopesticides, and organic farming, for that matter, come to the rescue (Vandamme et al., 2017). Soil microbiology studies soil microbes, their functions, and their effects on soil parameters (Tayyab et al., 2021). Bacteria, actinomycetes, fungus, algae, and protozoa are all soil microorganisms. Each group has distinguishing traits that differentiate them and their roles in the soil (Rao Subba et al., 1999). Biopesticide is a pesticide derived from a living organism or naturally occurring substances. Biochemical pesticides can also be produced from naturally occurring substances that can control pest populations in a non-toxic matter EPA, OCSPP, OPP, US (2015). The most commonly used microbe for microbial bio-pesticides is *Bacillus thuringiensis*, also known as Bt.

1.1.12.2 CHEMICAL APPLICATION

Microorganisms are particularly essential in industrial microbiology because they manufacture various products that are beneficial to humans and have dramatically transformed our lives and lifespans. Beverages, food additives, human and animal health goods, and biofuels are among these products

(Pandeeti et al., 2019). The utilization of complete microorganisms, their enzymes, or other enzymes expressed in microbes to manufacture products offers for easy access to building blocks and materials that are utilized as feedstocks or raw materials in industries like chemistry, food and feed, paper and pulp, textiles, and energy (Kumar et al., 2009). In the pre-treatment of lignocellulosic feedstock, microbes play a critical role. The input material is converted into a hydrolysate, which is a mixture of sugars and other products such as furans, organic acids, and phenolic compounds, by microbial enzymes (Olsson & Hagerdal, 1996). Microorganisms not only have a wide range of enzymes that enable a wide range of chemical conversions, but they also have a high metabolic activity that allows conversions to happen quickly (Kiruthika et al., 1990).

1.1.12.3 MEDICAL APPLICATION

The manufacture of novel pharmaceuticals generated in a specific organism for medical purposes is a medical application of industrial microbiology. Antibiotics are required for the treatment of a wide range of bacterial illnesses. Fermentation is used to manufacture several naturally occurring antibiotics and precursors. Vitamins are created in large numbers, either by fermentation or biotransformation (Erick, 2007). For example, Vitamin B 2 (riboflavin) is created in both ways. Biotransformation is most commonly utilized to make riboflavin, and glucose is the carbon supply for this reaction. A few microorganism strains were developed to boost the amount of riboflavin generated. *Ashbya gossypii* is the most common organism employed in this process (Chen Jiann-Shin et al., 2013) Industrial microbiology is the use of microbes on a large scale to synthesis commercially valuable products for a wide range of uses (Jit & Garg, 2015).

Microbial products can be broadly categorized into:

1. **Metabolic Production:** A metabolic pathway has been defined as converting a substrate into its final end-product (Newsholme et al., 1979). When dietary elements are transformed into metabolic products with a higher net anionic charge, hydrogen ions are released. Because it reflects the conversion of uncharged substrates into uncharged products, the oxidation of most carbohydrates or triglycerides to exhale CO_2 and water produces no acid (Mitchell et al., 1983). Metabolic engineering has been enabling the development of high-performance microbial strains to efficiently produce natural

and non-natural compounds from renewable non-food biomass (Huccetogullari et al., 2019).

2. **Biotransformation:** It is the process of converting organic molecules from one form to another to lessen their persistence and toxicity. A wide range of microorganisms and their products, such as bacteria, fungus, and enzymes, facilitate this process (Asha et al., 2009). Microbial biotransformation is widely employed in the transformation of a wide range of contaminants and chemicals, such as hydrocarbons, pharmaceuticals, and metals (Karigar et al., 2011).

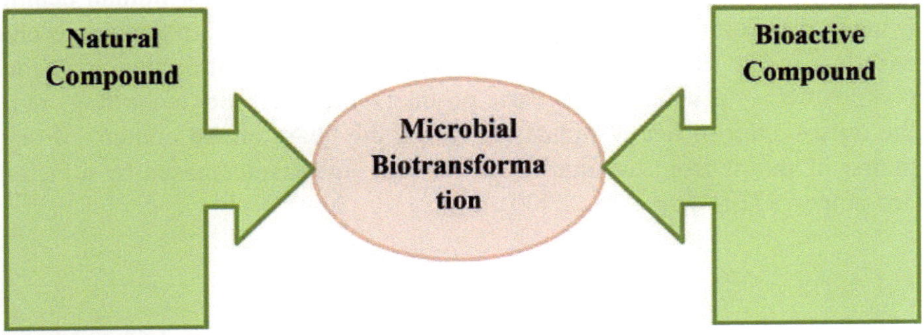

3. **Production of Biofuels:** Biofuel research aims at producing energy products such as alcohols (mainly ethanol, but also propanol and butanol, as well as propane and butane diols), diesel, hydrogen, and biogas from biological (mainly plant) sources (Mostafa, 2010). Plant biomass is a plentiful and renewable source of energy-rich carbohydrates that microorganisms can convert to biofuels quickly. Today, only bioethanol is manufactured on a large scale (Antoni et al., 2007). Despite growing environmental challenges, large volumes of fossil fuels are consumed daily. Biofuels created from various types of biomasses are used worldwide to preserve the environment and build a sustainable society (Sakuragi et al., 2011) (Table 1.4).

TABLE 1.4 Some of Biofuels and their Production Process

Biofuels	Process
Biomethanol	Thermochemical/microbial
Bioethanol	Microbial
Biobutanol	Microbial
Biohydrogen	Microbial
Biodiesel	Physical/chemical (enzymatic)

1.2 TREATMENT OF ORGANIC AND INDUSTRIAL WASTES

With the rapid increase of population and industrialization, the waste generation fact sheet in the current century usually varies country-wise (Salkin et al., 2003). Numerous industries (e.g., electroplating, metal-finishing, electronic, steel, and nonferrous processes, petrochemical, and pharmaceutical) discharge various toxic heavy metals into the environment. If improperly dealt with, these materials can threaten public health and the environment (Garcia-Valles et al., 2007). Most industrial wastes contain various metals like Zn, Cd, Cr, Cu, Ni, Pb, V, Mo, Co, which are toxic for human health if inhaled beyond the limited concentration (Marafi et al., 2008). In recent decades, many first-world countries have seen a dramatic change in the composition of urban organic waste products due to cleaner technology and the outsourcing of heavy industries. Thus, the heavy metal content of dry matter in sludge from Denmark (Zn, Pb, Cd, Hg) has decreased by a factor 3e5 since 1972 (Anon et al., 2006).

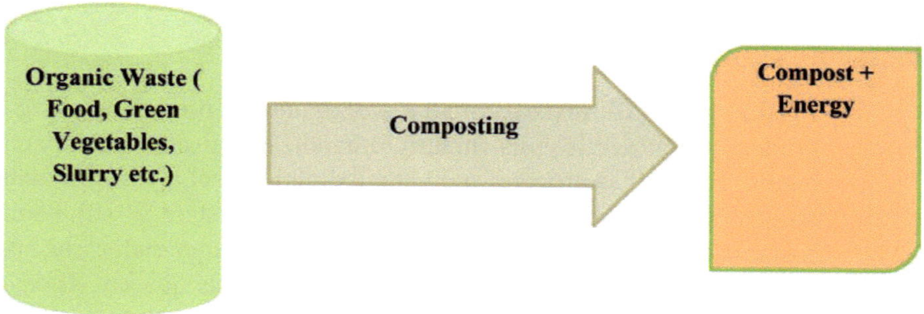

Excess activated sludge obtained from industrial waste treatment may contain contaminants that prevent it from being used as agricultural fertilizers. The choices for solid waste disposal from biological treatment plants have been significantly constrained due to recent regulatory amendments (McCann, 1998). Current waste treatment trends encourage waste recycling to the greatest extent possible to recover its material content and home and industrial waste treatment to utilize its energy content (Laca et al., 2011). Organic/degradable trash must be stabilized by stabilization before being dumped in an increasing number of landfills that only accept inert garbage. Because there is less possibility of gas and leachate formation, these landfills are low-emission landfills that require very little control and engineering than landfills that receive degradable trash (Matos et al., 2007).

1.2.1 PRODUCTION OF MICROBIAL BIOMASS

Biomass is a natural alternative for chemical and feedstock production with a short replacement cycle to fulfill the needs of the international fuel industry (Ahmed et al., 2005, 2007, 2009b). Corn stover and molasses can generate microbial biomass protein by sequential culture fermentation without costly pre-treatment or nutrient supplementation. The present results increase microbial biomass protein production information by sequential culture fermentation from waste products (Sibtain et al., 2010). In the soils treated with energy supplies, microbial biomass and biochemical activity tests increased, then dropped to the level of the control soil. The increases in biomass and enzyme activity that occurred at zero time due to adding exogenous microbes and enzymes to the ryegrass soil were not sustained after lengthy incubation (Nannipieri et al., 1993).

1.2.2 FOOD BIO-PRESERVATIVES

Food biopreservation refers to the use of natural or regulated microflora to increase the shelf life and improve the safety of food goods (Zinoviadou et al., 2016). Biocontrol or biopreservation refers to biological or regulated microbiota or antibacterial agents to extend the shelf life and safety of foods (Stiles et al., 1996). Biopreservation is the process of increasing the shelf life of food by utilizing natural resources such as flora and fauna, microbes, and/or their metabolites (Alexandru Mihai et al., 2018).

1.2.3 PRODUCTION OF STEROID

Steroid compounds represent a specific class of terpenoid lipids that contain in their structure gonane core of four cycloalkane rings (Baker et al., 2011), Steroids fulfill essential vital functions in the higher organisms: control various aspects of cell proliferation and differentiation, regulate signal transduction pathways by the binding to the respective intracellular receptors, some of them serve also as signaling molecules in cell-cell interactions, etc. (Bonnie et al., 2005). Construction and fermentation of engineered microorganisms open a new era in applied microbiology and industrial biotechnology of steroids. However, relatively low productivity, and sometimes, instability of the genetically modified strains, as well as scaling-up problems could be the reason why some promising laboratory bench developments, which are highly important for further practical use, still do not find wide commercial application (Donova et al., 2017). Steroid compounds are responsible for critical biological functions in cells: steroids of the estrane, androstane, and pregnane series show various hormonal activity; cardenolides are used for heart failure treatment, and bile acids are essential for lipid digestion and absorption. Sterols are crucial components of cell membranes and are precursors of bile acids and hormone steroids. They are required for cell stability, cell development, and proliferation (Sedlaczek et al., 1988).

Industrial microbiology incorporates the use of micro-organisms in food and other products. Microbial processes for industrial production of chemicals and various low-molecular weight compounds are quickly gaining practical importance for synthesizing high-purity products in an environmentally acceptable manner while realizing energy savings. To improve the productivity and yield of valuable chemicals, novel approaches must be developed. The application of natural products in food, pharmacy, medicine, and agriculture fields is extensively highlighted due to their chemical stability and biocompatibility. Microorganisms have been employed for several decades in large-scale production of various biochemicals, ranging from alcohol to antibiotics and processing foods and feeds.

KEYWORDS

- **agriculture application**
- **antimicrobial metabolite**
- **fermentation**
- **food bio-preservatives**
- **industrial enzyme**
- **microbial biomass**
- **microorganism**
- **organic and industrial wastes**

REFERENCES

Abbott, B. J., (1979). *Developments in Industrial Microbiology, 20,* 345.

Abdel-Aziz, S. M., Elsoud, M. M. A., & Anise, A. A. (2017). Microbial biosynthesis: a repertory of vital natural products. In *Food Biosynthesis* (pp. 25–54). Academic Press.

Ahmad, I., Khan, M. S. A., Aqil, F., & Singh, M., (2011). Microbial applications in agriculture and the environment: A broad perspective. *Microbes and Microbial Technology*.

Ahmed, S., Aslam, N., Latif, F., Rajoka, M. I., & Jamil, A., (2005). Molecular cloning of cellulase genes from *Trichoderma harzianum*. In: Atta-Ur-Rehman, Choudhary & Khan, (eds.), *Frontiers in Natural Product Chemistry* (Vol. 1, pp. 73–75).

Ahmed, S., Bashir, A., Saleem, H., Saadia, M., & Jamil, A., (2009). Production and purification of cellulose-degrading enzymes from a filamentous fungus *Trichoderma harzianum*. *Pak. J. Bot.*, *41*, 1411–1419.

Ahmed, S., Jabeen, A., & Jamil, A., (2007). Xylanase from *Trichoderma harzianum*: Enzyme characterization and gene isolation. *J. Chem. Soc. Pak.*, *29*, 176–182.

Anal, A., (2019). Quality ingredients and safety concerns for traditional fermented foods and beverages from Asia: A review. *Fermentation, 5*(1), 8.

Anandharaj, M., & Sivasankari, B., (2013). Isolation of potential probiotic *Lactobacillus* strains from human milk. *International Journal of Research in Pharmacy and Life Sciences,* 26–29.

Anil, K., & Nikita, C., (2017). *Role of Microbes in Dairy Industry*. Devi Ahilya University, India Submission.

Anon., (2011). *Vejledning og Gødsknings- og Harmoniregler.* Ministeriet for Fødevarer, Landbrug og Fiskeri, Plantedirektoratet, Denmark.

Antoni, D., Zverlov, V. V., & Schwarz, W. H., (2007). Biofuels from microbes. *Applied Microbiology and Biotechnology, 77*(1), 23–35.

Ardö, Y., (2006). Flavor formation by amino acid catabolism. *Biotechnol. Adv., 24*(2), 238–242.

Asha, S., & Vidyavathi, M., (2009). Cunninghamella--a microbial model for drug metabolism studies: A review. *Biotechnol. Adv.*, *27*(1), 16–29.

Baker, M. E., (2011). Origin and diversification of steroids: Co-evolution of enzymes and nuclear receptors. *Mol. Cell Endocrinol.*, *334*, 14–20.

Bisen, P. S., (2014). *Microbes in Practice* (pp. 1158–1224). New Delhi, IK International.

Bonnie, L., (2005). Bassler quorum sensing: Cell-to-cell communication in bacteria. *Annu. Rev. Cell Dev. Biol.*, *21*, 319–346.

Bounaix, M. S., Gabriel, V., Robert, H., Morel, S., Remaud, S. M., Gabriel, B., & Fontagné-Faucher, C., (2010). Characterization of glucan-producing *Leuconostoc* strains isolated from sourdough. *Int. J. Food Microbiol., 144*, 1–9.

Button, J. E., & Dutton, R. J., (2012). Cheese microbes. *Curr. Biol.*, *22*(15), 587–589.

Choi, J. M., Han, S. S., & Kim, H. S., (2015). Industrial applications of enzyme biocatalysis: Current status and future aspects. *Biotechnology Advances, 33*(7), 1443–1454.

Collado-Fernández, M., (2003). In: *Encyclopedia of Food Sciences and Nutrition* (2nd edn.).

Constantí, M., Poblet, M., Arola, L., Mas, A., & Guillamón, J. M., (1997). Analysis of yeasts populations during alcoholic fermentation in a newly established winery. *Am. J. Enol. Vitic., 48*, 339–344.

Dahal, N. R., Karki, T. B., Swamylingappa, B., Li, Q., & Gu, G., (2005). Traditional foods and beverages of Nepal: A review. *Food Reviews International, 21*(1), 1–25.

Danova, S., Petrov, K., Pavlov, P., & Petrova, P., (2005). Isolation and characterization of *Lactobacillus* strains involved in koumiss fermentation. *Intl. J. Dairy Technol., 58*(2), 100–105.

De Los Reyes-Gavilán, C. G., Fernández, M., Hudson, J. A., & Korpela, R., (2015). Role of microorganisms present in dairy fermented products in health and disease. *BioMed Research International, 2015*

Demain, A., (1981). Industrial microbiology. *Science, 214*(4524).

Demir, N., Bahc¸eci, K. S., & Acar, J., (2006). The effects of different initial *Lactobacillus plantarum* concentrations on some properties of fermented carrot juice. *Journal of Food Processing and Preservation, 30*(3), 352–363.

Dequin, S., Salmon, J. M., Nguyen, H., & Blondin, B., (2003). Wine yeasts. In: Boekhout, T., & Robert, V., (eds.), *Yeasts in Food* (pp. 389–412). Behr's Verlag, Hamburg.

Doelle, H. W., (1975). In: *Bacterial Metabolism* (2nd edn.).

Donova, M. V., (2017). Steroid bioconversions. *Methods in Molecular Biology*, 1–13.

Elshahed, M. S., (2010). Microbiological aspects of biofuel production: Current status, and future directions. *Journal of Advanced Research, 1*(2), 103–111.

Emain, A. L., (2010). *History of Industrial Biotechnology*.

EPA, OCSPP, OPP, US, (2015). *What are Biopesticides?* www.epa.gov (accessed on 18 November 2023).

Fernández-Lucas, J., Daniel, C., & Daniel, H., (2017). New trends for a classical enzyme: Papain, a biotechnological success story in the food industry. *Trends in Food Science & Technology, 68*, 91–110.

Garcia-Valles, M., Avila, G., Martinez, S., Terradas, R., & Nogues, J. M., (2008). Acoustic barriers obtained from industrial wastes. *Chemosphere, 72*, 1098–1102.

Godfrey, T., & West, S. I., (1996). Introduction to industrial enzymology. In: Godfrey, T., & West, S., (eds.), *Industrial Enzymology* (2nd edn., pp. 1–8). London: Macmillan Press.

Grumezescu, A. M., & Holban, A. M. (Eds.). (2018). *Food control and biosecurity* (Vol. 16). Academic Press.

Heller, K. J., (2001). Probiotic bacteria in fermented foods: Product characteristics and starter organisms. *American Journal of Clinical Nutrition, 73*(2), 374S–379S.

Huccetogullari, D., Luo, Z. W., & Lee, S. Y., (2019). Metabolic engineering of microorganisms for production of aromatic compounds. *Microb. Cell Fact., 18*, 41. https://doi.org/10.1186/s12934-019-1090-4.

Hunter-Cervera, J. C., & Belt, A., (1996). *Maintaining Cultures for Biotechnology and Industry*. Academic Press, New York, NY.

International Food Biotechnology Council, (1990). Chapter 4: Safety evaluation of foods and food ingredients derived from microorganisms. *Regul. Toxicol. Pharmacol., 12*, S114–S128.

Jiann-Shin, C., Zidwick, M. J., & Rogers, P., (2013). Organic acid and solvent production: Butanol, acetone, and isopropanol; 1,3- and 1,2-propanediol production; and 2,3-butanediol production. *The Prokaryotes*, 77–134.

Jit, S., & Garg, N., (2015). *Fermentation Process*.

Karigar, C. S., & Rao, S. S., (2011). Role of microbial enzymes in the bioremediation of pollutants: A review. *Enzyme Res., 2011*, 11.

Kirk, O., Borchert, T. V., & Fuglsang, C. C., (2002). Industrial enzyme applications. *Current Opinion in Biotechnology, 13*(4), 345–351.

Kiruthika, S., (1990). Role of microorganisms in industrial processes. *Microbiology*.

Kumar, A., Asthana, M., Gupta, A., Nigam, D., & Mahajan, S. (2018). Secondary metabolism and antimicrobial metabolites of Penicillium. In *New and Future Developments in Microbial Biotechnology and Bioengineering* (pp. 47–68). Elsevier.

Laca, A., Herrero, M., & Díaz, M., (2011). *Life Cycle Assessment in Biotechnology. Comprehensive Biotechnology,* 839–851.

Lisbeth, O., & Hahn-Hägerdal, B., (1996). Fermentation of lignocellulosic hydrolysates for ethanol production. *Enzyme and Microbial Technology, 18*(5), 312–331.

Lourens-Hattingh, A., & Viljoen, B. C., (2001). Yogurt as probiotic carrier food. *Intl. Dairy J., 11*(1), 1–17.

Maplestone, R. A., Stone, M. J., & Williams, D. H., (1992). The evolutionary role of secondary metabolites — A review. *Gene, 115*(1, 2), 151–157. doi: 10.1016/0378-1119(92)90553-2

Marafi, M., (2008). Stanislaus A. Spent catalyst waste management: A review part I-developments in hydroprocessing catalyst waste reduction and use. *Resources, Conservation and Recycling, 52,* 859–873.

Matos, S., & Hall, J., (2007). Integrating sustainable development in the supply chain: The case of life cycle assessment in oil and gas and agricultural biotechnology. *Journal of Operations Management, 25*, 1083–1102.

McCann, B., (1998). Disposable outcome. *Water, 19*, 11.

McFeeters, R., (2004). Fermentation microorganisms and flavor changes in fermented foods. *Journal of Food Science, 69*(1), FMS35–FMS37.

Michael, J. W., & Neil, L., (2013). *Morgan Are the Authors of Industrial Microbiology: An Introduction.* published by Wiley.

Mitchell, L. H., & Junjas, R. L., (1983). Metabolic production and renal disposal of hydrogen ions. *Kidney International, 24*, 709–713

Najafpour, G. D., (2007). Industrial microbiology. *Biochemical Engineering and Biotechnology*, 1–13. doi:10.1016/b978-044452845-2/50001-x.

Nannipieri, P., Muccini, L., & Ciardi, C., (1983). Microbial biomass and enzyme activities: Production and persistence. *Soil Biology and Biochemistry, 15*(6), 679–685.

Newsholme, E. A., & Crabtree, B., (1979). Theoretical pinciples in the approaches to control of metabolic pathways and their application to glycolysis in muscle. *J. Mol. Cell Cardiol., 11*, 839—856,

Okafor, N., (2007). *Modern Industrial Microbiology and Biotechnology*. Department of Biological Sciences, South Carolina, USA.

Panda, S. H., & Ray, R. C., (2007). Lactic acid fermentation of β-carotene rich sweet potato (*Ipomoea batatas* L.) into lacto-juice. *Plant Foods for Human Nutrition, 62*(2), 65–70.

Pandeeti, E. V. P., & Routhu, G. D., (2019). In: *Recent Developments in Applied Microbiology and Biochemistry.*

Pintado, M. E., Da Silva, J. A., Fernandes, P. B., Malcata, F. X., & Hogg, T. A., (1996). Microbiological and rheological studies on Portuguese kefir grains. *Intl. J. Food Sci. Technol., 31*(1), 15–26.

Prasanna, (2021). Industrially important microorganisms and their products of microbiology. *Journal of Food Microbiology*, 12, 13.

Rao, S., (1999). *Soil Microbiology* (4th edn.) Enfield: Science Publishers. Print.

Ronald, S. J., (2008). In: *Wine Science* (3rd edn.).

Sakuragi, H., Kuroda, K., & Ueda, M., (2011). Molecular breeding of advanced microorganisms for biofuel production. *Journal of Biomedicine and Biotechnology, 2011*, 1–11. doi: 10.1155/2011/416931

Salkin, F. I., (2003). Conventional and alternative technologies for the treatment of infectious waste. *Journal of Material Cycles Waste Management, 5*, 9–12.

Sanromán, M. A., & Deive, F. J., (2017). Food enzymes. *Current Developments in Biotechnology and Bioengineering* (pp. 119–142). Elsevier.

Sedlaczek, L., & Smith, L. L., (1988). *Biotransformations of Steroids. Critical Reviews in Biotechnology, 7*(3), 187–236.

Sibtain, A., Fayyaz, A., & Hashmi, A. S., (2010). Production of microbial biomass protein by sequential culture fermentation of *Arachniotus sp. & Candida utilis. Pak. J. Bot., 42*(2), 1225–1234.

Soukoulis, C., Panagiotidis, P., Koureli, R., & Tzia, C., (2007). Industrial yogurt manufacture: Monitoring of fermentation process and improvement of final product quality. *J. Dairy Sci., 90*(6), 2641–2654.

Stiles, M., (1996). Biopreservation by lactic acid bacteria. *Antonie Van Leeuwen, 70*, 331–345

Stone, M. J., & Williams, D. H., (1992). On the evolution of functional secondary metabolites (natural products). *Molecular Microbiology, 6*(1), 29–34.

Tayyab, M., Yang, Z., Zhang, C., Islam, W., Lin, W., & Zhang, H., (2021). Sugarcane monoculture drives microbial community composition, activity, and abundance of agricultural-related microorganisms. *Environmental Science and Pollution Research, 28*(35), 48080–48096.

Therdthai, N., (2014). *Fermentation. Bakery Products Science and Technology,* 325–334.

Vandamme, E. J., (2007). Production of vitamins, coenzymes and related biochemicals by biotechnological processes. *Journal of Chemical Technology & Biotechnology, 53*(4), 313–327.

Vieira, G. A., Souza, C. T., Silva, C. L., Mendonc¸a, B. F., & Parachin, N., (2018). Comparison of yeasts as hosts for recombinant protein production. *Microorganisms, 6*(2), 38.

Witthuhn, R. C., Schoeman, T., & Britz, T. J., (2005). Characterization of the microbial population at different stages of Kefir production and Kefir grain mass cultivation. *Intl Dairy J, 15*(4), 383–389.

Witthuhn, R. C., Schoeman, T., & Britz, T. J., (2005). Characterization of the microbial population at different stages of Kefir production and Kefir grain mass cultivation. *Intl. Dairy J., 15*(4), 383–389.

Yusuf, M. (2018). Natural antimicrobial agents for food biopreservation. In *Food Packaging and Preservation* (pp. 409–438). Academic Press.

Yvon, M., & Rijnen, L., (2001). Cheese flavor formation by amino acid catabolism. *Intl. Dairy J., 11*(4), 185–201.

Zinoviadou, K. G., Gougouli, M., & Biliaderis, C. G., (2016). Innovative biobased materials for packaging sustainability. *Innovation Strategies in the Food Industry,* 167–189.

CHAPTER 2

History of Industrial Microbiology

2.1 THE ANCIENT ERA

The branch of industrial microbiology has its roots deep in ancient times. Long before their "discovery," microorganisms were used in various ways to serve the needs and desires of humans, such as milk, fruits, and vegetable preservation, and also to enhance and improve the quality of life by producing various products like yogurt, cheeses, bread, pickled foods, and vinegar. Since ancient days, yeast has been used in fermentation. The oldest fermentation known, the conversion of sugar to alcohol by yeasts, was used to make beer in Samaria and Babylonia as early as 7000 BC. In broad terms, fermentation involves using microorganisms to carry out enzyme-catalyzed transformations of organic matter. Fermentation has been performed as an art for many centuries. Wine-making is thought to have been practiced at least since 10,000 BC. Historians believe that the Egyptians produced beer in 5,000–6,000 BC. Barley, germinated in earthenware vessels, was crushed, mixed as dough and baked, and finally soaked in water and allowed to produce beer. About 4000 BC, the Egyptians used brewer's yeast for carbon dioxide (CO_2) production in the leavening of bread.

Records of the transformation of milk into products are there such as cheese, date back to 5000 BC when it was observed that milk carried in calf stomachs (which contain milk-clotting enzymes) tended to curdle. For preservation, milk was fermented to lactic acid (LA) to make yogurt and converted into kefir and koumiss using *Kluyveromyces* species in Asia.

Algae of the genus *Spirulina* were harvested from alkaline ponds by the ancient Aztecs in Mexico. The origins of a various indigenous fermented foods in the Orient and elsewhere, now known to be based on fermentation, enzyme production, and enzymatic hydrolysis using surface culture methods, clearly also go back many thousands of years. The earliest references to

Handbook of Industrial Food Microbiology. Manju Nehra & Vikash Nain
© 2025 Apple Academic Press, Inc. Co-published with CRC Press (Taylor & Francis)

distilled potable spirits date to 1000 BC in China. By the 14th century AD, the distillation of alcoholic spirits from fermented grain, a practice which was supposed to be originated in China or in the Middle East, was standard in many parts of the world. Vinegar has probably been known for as long as wine-making has been practiced, although the earliest records referring to vinegar are in the Old and New Testaments. Vinegar manufacturing began in France at the end of the 14th century in Orlean region, and the technique used is known as the Orleans method.

Production of modified foods and beverages involving fermentation has been operated for approximately 10,000 years before microorganisms were recognized. There is also evidence that, gradually, process improvements were made in these traditional technologies. Microscopic examination of sediments of excavated beer urns dating from 3400 BC to 1440 BC almost certainly indicates yeast cells. It has been claimed that the cast in the more recent sediment was of improved purity. In the 17th century, Antonie van Leeuwenhoek, a Dutch merchant with no university training but a keen amateur interest in the construction of microscopes, turned his simple lens to examining water, decaying matter, and scrapings from his teeth.

On the pharmaceutical side, moldy cheese, meat, and bread had been employed in folk medicine for thousands of years to heal injuries/wounds and treat infections. In retrospect, the beneficial effect of such treatments was undoubtedly due to their antibiotic activity.

2.2 MODERN MICROBIAL TECHNOLOGY

In 1680, Antonie van Leeuwenhoek, a pioneering Dutch microscopist, studied yeast cells for the first time when he used a primitive microscope to examine a drop of fermented beer. However, his discovery of the yeast cell was soon forgotten, and fermentation continued, as it has been before, to be almost exclusively studied by chemists who did not consider fermentation to involve living material. Then, in 1836–1837, three scholars, Cagniard-Latour, Schwann, and Kutzing, all independently considered yeast to be a living 'thing.' Berzelius, Wohler, von Liebig, and others debunked these findings.

In 1847, Blondeau, a physics professor, studied fermentations involving acetic acid, LA, urea, and butyric acid, and seems to have been the first one who stated that different fermentations were carried out by different 'fungi.' Despite these findings, it was not until 1856–1857 that Louis Pasteur, after conducting extensive research on beer and wine fermentations, concluded that when yeast cells are compelled to work in the absence of oxygen, they ferment sugar into ethanol and CO_2.

Pasteur also examined a variety of other fermentations. He discovered that a racemic mixture of d and l-tartrate is produced by an organism (possibly a *Penicillium* sp.) on selective fermentation of ammonium tartrate. He studied butyric acid synthesis by cylindrical species under anaerobic conditions and acetic acid fermentation. Pasteur and several other investigators had discovered the antagonistic actions of one microbe on another in the 1870s and anticipated the medicinal applications of these microorganisms. Another significant landmark in industrial and medical microbiology history occurred in 1881 when medical doctor Robert Koch initiated the development of pure culture techniques and other traditional bacteriological approaches.

The ancient art of utilizing yeasts and molds for food and beverage processing and the pioneering microbiological experiments of scientists like Pasteur and Koch laid the foundation for the new period of industrial fermentation technology. The technology for surface or semi-solid fermentation processes had developed through oriental foods. As seen in soy sauce manufacturing, the molds' potential to hydrolyze starch and protein opened the way for industrial enzyme production. Industrial procedures for processing microbial food protein (single-cell protein (SCP)) and microbial inoculants were made possible by cultivating algae and yeast. The traditional production of vinegar and Pasteur's more recent studies on the production of acids prepared the path for developing fermentation to produce organic acids. Koch's invention of pure culture methodology equipped others with the tools to research specific microbial strains for industrial applications. It paved the way for the usage of pure cultures, sterilized media, and aseptic environments in industrial fermentations.

2.2.1 SIGNIFICANT MILESTONES IN MODERN INDUSTRIAL MICROBIOLOGY

- Anthony Leeuwenhoek published his first observations of microorganisms in 1677.
- The experiments of Spallanzani in 1799 and Schwan in 1837 disproved the idea of spontaneous microorganism generation and provided a method for sterilizing liquids (by heat) and air (by heat).

- Schwan's findings also suggested that alcoholic fermentation was caused by a fungus or mold, such as yeast, and inoculation accelerated fermentation.
- However, Luis Pasteur's studies on lactic acid fermentation, which included microscopic features of the microorganisms and a suitable medium for the process, are widely regarded as the beginning of microbiology. This chapter is widely considered to be the scientific foundation of industrial microbiology.
- Pasteur published the first synthetic medium for microorganisms in 1860, and he used it to study the fermentation of alcohol.
- In 1861, Luis Pasteur discovered that the existence or absence of CO_2 affects yeast growth and physiology (and therefore the accumulation of the fermentation component, alcohol). The Pasteur phenomenon is a principle that applies to other microorganisms as well.
- In 1878, Lister identified the dilution method for obtaining the first pure lactic acid bacterium microbial culture.
- In 1881, Robert Koch identified a simplified and more reliable method for extracting pure cultures from separate isolated colonies formed on solidified medium; this method is still commonly used today.
- In 1876, Cohn demonstrated that bacterial spores are heat resistant and invented the 'intermittent sterilization' technique for their inactivation.
- In 1897, Buchner showed alcohol fermentation using cell-free yeast juice, claiming that fermentation was caused by a proteinaceous enzyme.
- In 1901, Wildiers discovered that yeast needed growth factors (vitamins) to develop, especially at low inoculum levels; vitamins are still used in fermentation today.
- In 1929, Alexander Fleming discovered penicillin developed by Penicillium developing as a contaminant in a Staphylococcus Petri plate by mistake. Fleming pioneered bacteria to evaluate penicillin's antibacterial efficacy, demonstrating its low toxicity to humans and animals. During WWII, an ambitious hunt for antibiotics led to the discovery of streptomycin, chloramphenicol, tetracyclines, and other antibiotics. Later research has shown to metabolically blocked microorganisms that store vast quantities of metabolic intermediates.

2.3 PLANT AND ANIMAL CELL CULTURING

Although methods for developing mammalian cells *in vitro* have been used for over a century, advancements are much slower in this stream than those in microbial science. In 1885, Roux used living tissue from a chick embryo to conduct the first "explanation" research. Many of the classic mammalian cell-culture strategies were invented by Harrison and Carrel around 1910. They improved the salt and amino acid composition of growth media, established bioreactors for the propagation of viruses and other products, and, for the first time, used supplementary plasma and clotted lymph as growth-promoting additives. Morgan and his collaborators produced the first chemically defined medium suitable for maintaining chick embryo growth for 4–5 weeks in 1950. Most existing strain-specific cell-culture formulations are based on this team's specific vitamin and amino acid specifications for reproducible L-cell development. From the 1950s onwards, these advancements enabled viral vaccines to be developed utilizing cell-culture techniques, paving the way for a significant expansion in the industrial use of cell culture for the production of monoclonal antibodies and other items in the mid-1980s.

Gautheret was the first to successfully culture plant cells in 1934. Plant cell-culture methods had progressed to the point that they could be used for strain production in horticulture and agriculture in the 1950s and 1960s. In the late 1950s and early 1960s, large-scale tobacco and vegetable cell production was investigated. The use of plant cells in culture to generate high-value secondary metabolite plant products gained significant interest. Mitsui Petrochemical Company in Japan began commercial development of shikonin from *Lithospermum erythrorhizun* in 1985, marking the first time a secondary plant metabolite was commercially produced using cell culture.

2.4 FERMENTED FOOD

Although modern industrial food fermentations evolved from conventional fermented food processes, food fermentations have continued to grow technologically. Current commercial demands of large-scale processing, high and reliable quality, low prices, and product diversity have led to the development of bread baking, cheese manufacturing, brewing, and distilling processes. Modern baking and dairy fermentation processes have increased performance, automation, and quality control due to the highly standardized processing of baker's yeast and dairy starter cultures and the use of industrial enzymes.

The usage of a significant amount of yeast at higher temperatures has accelerated the fermentation of bread dough. The use of microbial amylases liberates fermentable sugars from starch grains, supplying sugar for the yeast to ferment into CO_2 bubbles, which expand the dough and provide it with its unique texture. Microbial protectants are used to partly hydrolyze wheat gluten proteins in bread, thus enhancing dough handling properties, loaf volume, and shape.

One of the variables that have led to the production of a diverse range of high-quality cheese products is the application of cheese starter cultures. The production of cheese is no longer reliant on a sudden milk infection. The usage of starter cultures and other process assistance, in collaboration with new milk-handling machinery, has automated processes. There were severe issues with bacteriophage contamination in starter cultures. Researchers used starter culture mixtures, rotated phage-unrelated starter strains, and phage inhibitors to resolve these concerns. Bacteriophage-insensitive strains have been successfully produced since the early 1980s.

Many advances have happened in the alcoholic liquor industry since Pasteur's era. Due to the pasteurization process, the beer market increased from a localized sector to a regional and global scale. The brewing, distilling, and wine-making industries have rapidly evolved from small craft-like businesses to massive complexes to maintain a standardized commodity despite changing operations' raw materials, crop, and size. Fermentations are standardized by regulating parameters such as inoculation or pitching intensity, yeast-cell viability, yeast storage conditions, dissolved oxygen concentration at pitching, soluble nitrogen, fermentable sugar concentrations in wort, and fermentation temperature to produce a reproducible product. Fermentation innovations such as continuous fermentation processes, brewing at high wort gravity, use of un-malted cereals and microbial enzymes and production of low carbohydrate beer have been used with varying levels of success. Traditional genetic approaches have been used to enhance beneficial yeast properties and remove undesirable traits.

2.5 SINGLE-CELL PROTEIN (SCP) AND BAKER'S YEAST PRODUCTION

During Second World War in Germany, baker's yeast was created on molasses and used as a protein alternative for human consumption. SCP techniques were utilized for the first time in the history of modern fermentation. Candida utilis was later employed as a protein source for animals and humans in Germany, having been cultured on sulfite waste liquor from pulp and paper manufacturing and sugar derived from acid hydrolysis of wood. Since then, the procedure has been used in the United States, Switzerland, Taiwan, and the Soviet Union. Several corporations developed SCP plants in Europe, the United States, and Japan from 1968 onwards. Some of these plants have closed due to regulatory compliance or high manufacturing prices, although others continue their research and development efforts. In 1981, 1CI in England increased its SCP output by 3,000 t/month of Pruteen for animal feed, utilizing the aerobic bacterium *Methylophilut methylotrophus*. Mycoprotein, the first novel microbial protein developed through fermentation to receive government permission for human use, was produced by Rank Hovis McDougall in England using the fungus *Fusary graminearum*. Mycoprotein's filamentous morphology gives it a natural texture identical to beef, allowing it to demand a reasonable market price and achieve consumer appeal.

In the 17th century, commercial bakers acquired their yeast from local breweries. Brewer yeast was eventually replaced by distiller's yeast, which was then replaced by baker's yeast, as it imparts bitter flavor and has variable fermentation activities. Pressed baker's yeast was first developed in 1781 using the Dutch method. Later in 1846, the Vienna process produced low yeast yields of 5% and 14%, respectively, based on raw material, and concurrent spirits yields of about 30%. Between 1879 and 1919, remarkable progress has been achieved in commercial yeast biomass fermentations, resulting in a baker's yeast industry independent from alcoholic liquor manufacturing. Marquardt implemented aeration of the grain mash in 1879, which improved yield while reducing associated spirit production by 20%. Further refinements involving gradual sugar feeding, the first example of a fed-batch fermentation technique, increased processing productivity near the theoretical optimum without causing alcohol formation.

2.6 FOOD ADDITIVES

2.6.1 ORGANIC ACIDS

Fermentation of LA for commercial purposes started in 1881. *Lactobacillus delbruckii* is used to manufacture 50% of industrial LA. Initially, Citric acid was manufactured by extraction from lemon juice, but these days, it is produced from microbial fermentation of glycerol and other chemicals.

Citrate was first developed through industrial fermentation in 1923. After 10 years, annual global production had crossed 10,000 t, with fermentation contributing for more than 80% of the total. Current industry estimates the annual production at more than 3,50,000 I, mostly manufactured exclusively by fermentation. Surface culture methods were initially used with *Aspergillus niger* as culture organism. After Second World War, submerged *Aspergillus niger* processes were developed, and around 1977, a more effective submerged mechanism utilizing *Candida* species was commercialized. Gluconic and itaconic acids are two other organic acids that have been generated economically by fermentation. Food-grade acetic acid (vinegar) is made solely by the oxidation of ethanol by *Acetobacter aceti*, although even today, industrial-grade acetic acid is made mainly through chemical processes.

2.6.2 VITAMINS

Chemical processes are currently used to manufacture the most of vitamins. While fermentation methods for various group B vitamins, including thiamine biotin, folic acid, pantothenic acid, pyridoxine, vitamin B_{12}, and riboflavin, have been established, only the last two, i.e., B_{12} and riboflavin are developed by microbial application. The chemical production of vitamin B_{12} is highly complex, and it is exclusively commercially manufactured employing Pseudomonas strains in industrial fermentation. Fermentation techniques remain competitive with chemical synthesis or semi-synthesis approaches in riboflavin manufacturing. Fermentation is accountable for 30% of the global riboflavin supply. *Eremothecium ashbyii* was first used to manufacture the vitamin in 1935, and yields were eventually reported high up to 5.3 g/1. A unique *Ascomycete* genus, *Ashbya gossypii*, later became the favored strain due to strain instability. A patent from 1984 mentions a *Bacillus subtilis* recombinant strain was able to produce 1.5 g riboflavin in a 24-hour fermentation, which is significantly faster (almost five times) than *Ascomycetes* process.

2.6.3 AMINO ACIDS

Aerobic fermentation synthesis of amino acids has increased dramatically over the last 30 years. Monosodium glutamate and lysine are the most abundant amino acids, with annual global production of more than 3.5 million and 2 million tons, respectively. Fermentation processes are the outcome of comprehensive studies into the biochemical pathways that regulate microbial biosynthesis of amino acid. Combining mutations and fermentation process regulation has resulted in amino acid overproduction.

The techniques used aim to stimulate cell uptake of new materials, prevent or hinder side reactions, induce, and activate biosynthetic enzymes, reduce and inhibit enzyme activity that would degrade the product, and facilitate excretion of the amino acid. The traditional research method used to produce glutamic acid fermentation has aided the advancement of fermentation processes to manufacture other primary metabolites. Consequently, nearly all amino acids are now commercially manufactured by fermentation. The flavor enhancers inosine and guanosine monophosphates were produced using similar techniques.

2.6.4 BIOPOLYMERS

For the development of microbial biopolymers, a variety of industrial fermentation techniques have been produced in recent years. Most polymers are synthetic and sensitive to the cost of petroleum feed stocks so that biopolymers will be especially important if oil prices go up. Xanthan gum, used as a suspension stabilizer and gelling agent, is the most effective biopolymer in the fermentation efficiency rate. Other microbial gums of commercial interest are produced by *Azotobacter bindandii* (alginate), *Aureobasidium pullulans* (pullulan), *Scaretium specics* (scleroglucan) and *Pseudomonas elodea* (gellan). Extracellular polysaccharide gums' high molecular weight and viscosity pose challenges for fermenter mixing, mass and heat transfer, which must be considered when developing an ideal fermenter design for these processes. Poly-hydroxybutyrate, a thermoplastic polyester accumulated intracellularly in certain species of *Alcaligenes eutrophus* to a level of up to 70% of biomass weight, is a promising polymer currently being produced. Biodegradable plastic will be able to substitute non-biodegradable petroleum-based plastics with such products.

2.7 CHEMICALS

2.7.1 ALCOHOLS AND KETONES

During World War I, deliveries of imported vegetable oils, used for the manufacturing of glycerol, were cut off from Germany. A fermentation method was rapidly invented to manufacture glycerol for use in explosives manufacture. The method was founded on the discovery by Neuberg that glycerol is formed by yeast, at the cost of ethanol, in the presence of sodium bisulfite. The British fulfilled a war need for acetone by inventing the anaerobic acetone-butanol fermentation involving *Clostridium acetobutylicum.* This was the first large-scale fermentation utilizing pure-culture techniques to avoid contamination. After World War I, demand for acetone decreased. Still, demand for n-butanol increased, and fermentation was used to manufacture it until the 1950s, when the price of petrochemicals fell below that of starch and sugar-based substrates. National Chemical Products, South Africa, conducted the acetone-butanol fermentation commercially until recently, when petroleum was scarce due to worldwide embargoes.

Since the late 1800s, chemically manufactured synthetic alcohol has been widely used in the United States. With repeal of prohibition of alcohol in the United States, the industrial alcohol fermentation industry has grown to an account for 77% of the industrial alcohol market in 1941. During the 1950s and 1960s, ethylene was inexpensive, and powerful hydrogen processes for transforming it to ethanol were produced, rendering synthetic ethanol more cost-effective than fermentation methods. The 1973 oil embargoes and the resulting spikes in oil prices sparked global interest in alternative green fuel systems. Brazil initiated a national alcohol program in 1987 intending to increase fermentation capability to generate 3 billion gallons of ethanol each year from cane sugar.

In the Gasohol Program, the United States government promoted the development of fermentation-based alcohol production from corn. The excise tax on gasoline (which contains 10% alcohol) was decreased, and a significant number of small and large-scale fermentation plants were developed to manufacture alcohol using batch and continuous fermentation methods. By the mid-1980s, the Program aimed to produce an average capacity of 1.8 billion gallons. While the amount of ethanol produced by *Kluyveromyces fragilis* fermentation of whey is small relative to the volume produced by *S. cerevisiae* from corn or cane-sugar, it does contain ethanol. Whey fermentations must occur beside the cheese factory since the expense of transporting whey, which comprises just around 5% w/v lactose, is prohibitive.

Fermentation is currently used to manufacture only a limited amount of industrial chemicals. Petroleum and natural gas are used to make almost all commodity chemicals. The fluctuating cost of fuel, the volatile availability of petroleum, and worries regarding non-renewable resource depletion have intensified interest in non-petroleum feedstocks for chemical and oil development.

2.7.2 MICROBIAL INSECTICIDES

This century, chemical insecticides have been widely utilized in agriculture and public health applications with great success. However, criticisms that they may kill non-target organisms or that target organisms may become resistant have led to the development and commercialization of microbial insecticides and on-going research is continuing in this area. Production of *Bacillus thuringiensis* far exceeds any other commercially produced microbial insecticide. Ideal fermentation parameters are maintained to maximize yield and bioactivity of the crystal endotoxin that produced concomitantly on sporulation. Different strains of *B. thuringiensis* and other bacterial and fungal insect pathogens exhibit different toxicities. Toxin-producing genes from *Bacillus thuringiensis* have been cloned into *E. coli* and *Bacillus thuringiensis subtilis*, which provides the possibility of increasing growth rates and to change the targeted spectrum.

2.7.3 GIBBERELLINS

Gibberellins a diterpenoid compounds with four cyclic carbon rings, are used to speed up barley germination in the malt industry. In 1938, the first gibberellin was discovered. There are currently 60 different substances believed to exist. Surface culture was used to manufacture gibberellic acid in the past, and it took a prolonged fermentation to get yields of 40–60 mg/l. Submerged culture is now used in commercial fermentations to produce considerably higher yields.

2.8 MICROBIAL ENZYMES

The first commercial application of isolated microbial enzymes goes back to Jokichi Takamine, a Japanese immigrant to the United States who patented a technique for extracting diastatic enzymes from molds in 1894, which was sold as Takadiastase. His technique, which included the mold growing on the surface of a solid substrate like wheat or bran, certainly represented his understanding of similar methods used to prepare oriental fermented foods. Boidin and Effront, from France and Germany, respectively, pioneered the invention of bacterial industrial enzyme fermentation with their 1917 patent, which outlined the use of *Bacillus subtilis* and *Bacillus mesentericus* to manufacture amylases and diastases utilizing surface cultivation techniques.

Surface culture methods for developing fungal enzymes were used in the United States well into the 1950s and are still used today, especially in Japan. Submerged fungal fermentation for enzyme production was produced in the United States and Europe due to the submerged fermentation experience obtained during establishing the penicillin production method at the Northern Regional Research Laboratories (NRRL) in the United States.

The commercialization of industrial microbial enzymes laid the foundation for the advancement of enzyme application technology, especially in food processing areas like brewing, fruit juice production, and hydrolyzed starch syrup production. Following alkaline proteases in washing powder formulations in the 1960s, the demand for synthetic enzymes exploded. The synthesis and application of microbial glucose isomerase for the production of high-fructose corn syrup was a significant development in the 1970s. This significantly influenced the nutritional sweetener industry, previously driven by sucrose. The carbonated soft drink sector is the primary commercial user of sugar in the United States. By 1984, all big soft drink producers agreed to replace sucrose entirely with HFCS. The industrial processing of high-temperature stable α-amylase from *Bacillus licheniformis* was another important technological breakthrough in the 1970s. The capability of this enzyme to liquefy starch at temperatures as high as 110°C has accelerated starch hydrolysis processes while also triggering researchers to isolate or develop other enzymes with high stability.

2.9 HEALTHCARE PRODUCTS

2.9.1 ANTIBIOTICS

Following Pasteur's hypothesis that the antagonistic results of one microbe or another may have therapeutic value, numerous microbial formulations were evaluated as medicines for 50 years with little success. Finally, in 1928, Alexander Fleming discovered that when *Penicillium notatum* was present as a contaminant in *Staphylococcus aureus* culture dishes, it destroyed the bacteria. He demonstrated that the active ingredient, penicillin, could kill a wide range of bacteria. After isolation of penicillin in a stable active form by Florey and Chain around 1940 remarkable antibacterial activity was demonstrated. A commercial fermentation was developed with the assistance of the NRRL in the USA and the involvement of several US pharmaceutical companies. As a result of the rapid production of penicillin fermentation, the antibiotic industry has emerged. Soon afterwards several new antibiotics, including streptomycin, were isolated from *Streptomyces* species and *cephalosporins* were produced by *Cepihalosporiumm acremontum*. While the *Penicillins*, cephalosporins, and sucptomycins were the most important early antibiotic discoveries, hundreds of new antibiotics have been isolated annually from the 1940s right up to the present day.

2.9.2 STEROID

Using microorganisms to carry out extremely complex and selective enzymatic transformation reactions of pharmaceuticals was a significant step in producing steroid hormonal drugs. Cortisone, a steroid secreted by the adrenal gland, was discovered to reduce the pain in rheumatoid arthritis patients in the 1940s. A 37-step chemical synthesis technique was produced, resulting in cortisone development for USD 200 per gram.

The Upjohn Company discovered a *Ruzopus arrhizus* that could hydroxylate progesterone to manufacture I1-a-hydroxyprogesterone in 1952. This enabled the cortisone synthesis process to be reduced from 37 to 11 stages, lowering the cost to USD 16 per gram. Over the years, process improvements have been made to reduce the manufacturing cost to reduce cortisone cost to USD 0.46/g in 1980. Some steroid medications, such as aldosterone, prednisone, and prednisolone, have been synthesized using similar microbial biotransformation techniques. Many medicinal and non-medical chemicals are synthesized using these strategies.

2.9.3 VACCINES

In ancient Hindu texts, infectious diseases are listed, and tuberculosis lesions have been discovered in mummified remains taken from early Egyptian tombs. Jenner discovered that individuals inoculated with cowpox or 'vaccinia' were resistant to subsequent challenges with smallpox in 1798 and identified the capacity to confer disease tolerance by vaccination. Behring and Kitasato became the first to immunize livestock with inactivated diphtheria and tetanus toxins in 1890. Following Koch's discovery of *Vibrio cholerae*, Ferran Y. Clua, a Spanish surgeon, delivered the first cholera vaccine in 1885 to over 30,000 Spaniards through subcutaneous injection of live broth cultures. During World War I, typhus vaccination programs were implemented in the armies, and similar tetanus vaccination programs were enacted during World

War II. The use of the diphtheria vaccine in mass vaccines began in the late 1930s. Following studies on the efficacy of experimental trials of these vaccinations in 1955, pertussis and poliomyelitis vaccines were widely used. Live attenuated strains of disease-carrying pathogens or closely associated strains, inactivated pathogenic strains, or cell components bearing antigens that successfully trigger immunity to targeted infectious diseases are used in bacterial vaccines today. Although vaccine development is a practical application of fermentation, it is achieved on a limited scale. Fermentation technology contains batch or continuous operations, with conditions customized to optimize cellular growth for immunogenic products.

For several years, viral vaccine technology was primitive due to the requirement for animals as virus sources. The use of chick embryos as a source of viruses was developed, and the tissue culture cells technique was adopted around 1950. Viral vaccine manufacturing reached the industrial fermentation arena with cell-culture techniques. Mammalian cells are cultured in an appropriate medium, and then at a particular stage of growth, the culture is inoculated with the virus, which multiplies in the mammalian cell host. Cell culture enables large quantities of virus to be prepared with much less contamination by extraneous material from host tissue, contaminating bacteria or viruses than earlier methods.

2.10 MODERN ADVANCEMENT IN INDUSTRIAL MICROBIOLOGY

2.10.1 MONOCLONAL ANTIBODIES

Kohler and Milstein fused mouse myeloma (skin cancer cell) with an antibody-producing white blood cell in 1975 to create a hybrid (hybridoma) cell that merged the parent cells' distinct cell-dividing and antibody-producing capacities. They invented the basic technology for producing monoclonal antibodies, preparations of individual-specific antibodies obtained from cells derived from a single ancestor or clone. Monoclonal antibodies have high binding specificity for individual receptors on a molecule or surface. Many existing and potential applications take advantage of this unique characteristic, particularly in clinical analysis and disease therapy. Monoclonal antibody production has been carried out *in vivo* by injection the hybridoma clone into the abdominal cavity fluid of the animal (usual mouse) or *in vitro* in various cell-culture systems. Recent developments in hybridoma technology and genetic engineering have extended the scope and potential of industrial fermentation technology.

2.10.2 RECOMBINANT DNA TECHNOLOGY

By putting foreign genes created from outside cell into a host organism where they do not naturally occur, genetic engineering allows for the creation of combinations of heritable material. In 1973, the first gene was cloned. Since then, techniques have been developing that can currently or will soon enable 'foreign' proteins to be produced commercially by various recombinant strains of prokaryotes and eukaryotes, including animal and plant cells. The pharmaceutical industry has invested hugely in this R&D activity. The first therapeutic product to achieve regulatory approval was human insulin, produced by a recombinant *E. coli*. Various other proteins have been produced in recombinant strains, including other human hormones and growth factors, anti-tumor and anti-viral compounds, viral antigens, and enzymes. The first genetically engineered vaccine to be marketed as a veterinary vaccine for pseudorabies, a herpes virus that infects swine, was introduced in 1986.

Initial pioneering genetic engineering research and the first products commercialized used *E. coli* as host since its genetic systems were more fully understood. Workhorses of the fermentation industry, *Bacillus, Aspergillus,* and *Saccharomyces* spp. maybe better hosts in the longer term once their genetic systems are more fully understood. Proteins and peptides, the translation products of structural genes, are the first obvious targets for production using recombinant DNA technology. In the longer-term, it may be possible to manipulate primary and secondary metabolism to improve metabolite production by genetic engineering of key enzymes in the metabolic sequence.

KEYWORDS

- **antibiotics**
- **gibberellins**
- **healthcare products**
- **industrial microbiology**
- **microbial insecticides**
- **monoclonal antibodies**
- **recombinant DNA technology**

REFERENCES

AL, D., (1957). Inhibition of penicillin formation by lysine. *Archives of Biochemistry and Biophysics, 67*(1), 244–246.

Ames, B. N., & Martin, R. G., (1964). Biochemical aspects of genetics: The operon. *Annual Review of Biochemistry, 33*(1), 235–258.

Arnstein, H. R. V., & Morris, D., (1960). The structure of a peptide, containing α-aminoadipic acid, cystine and valine, present in the mycelium of *Penicillium chrysogenum. Biochemical Journal, 76*(2), 357.

Bagdasarian, M., (1982). *Host: Vector Systems for Gene Cloning in Pseudomonas. 16*(1–3), 237–247.

Borneman, A. R., Schmidt, S. A., & Pretorius, I. S., (2013). At the cutting edge of grape and wine biotechnology. *Trends in Genetics, 29*(4), 263–271.

Brotzu, G. (1948). *Ricerche su di un Nuovo Antibiotic* (pp. 1–11). Lavori Dell'istituto d'Igiene di Cagliari.

Davey, V. F., & Johnson, M. J., (1953). Penicillin production in corn steep media with continuous carbohydrate addition. *Applied Microbiology, 1*(4), 208.

Demain, A. L., & Masurekar, P. S., (1974). Lysine inhibition of in vivo homocitrate synthesis in *Penicillium chrysogenum. Microbiology, 82*(1), 143–151.

Dixon, B., (2006). Sulfa's true significance. *Microbe, 1*, 500, 501.

Fleming, A., (1929). On the antibacterial action of cultures of a penicillium, with special reference to their use in the isolation of B. influenzae. *British Journal of Experimental Pathology, 10*(3), 226.

Gal, J., (2008). The discovery of biological enantioselectivity: Louis Pasteur and the fermentation of tartaric acid, 1857—A review and analysis 150 yr later. *Chirality: The Pharmacological, Biological, and Chemical Consequences of Molecular Asymmetry, 20*(1), 5–19.

Gottshall, R. Y., Roberts, J. M., Portwood, L. M., & Jennings, J. C., (1951). Synnematin, an antibiotic produced by Tilachlidium. *Proceedings of the Society for Experimental Biology and Medicine, 76*(2), 307–311.

Holloway, B. W., (1969). Genetics of *Pseudomonas. Bacteriological Reviews, 33*(3), 419.

Jarvis, F. G., & Johnson, M. J., (1947). The role of the constituents of synthetic media for penicillin production. *Journal of the American Chemical Society, 69*(12), 3010–3017.

Johnson, I. S., (1983). Human insulin from recombinant DNA technology. *Science, 219*(4585), 632–637.

Johnson, M. J., (1952). Recent advances in penicillin fermentation. *Bulletin of the World Health Organization, 6*(1, 2), 99.

Kuninaka, A., (1964). History and development of flavor nucleotides. *Food Technol., 18*, 287–293.

Linko, Y. Y., Javanainen, P., & Linko, S., (1997). Biotechnology of bread baking. *Trends in Food Science & Technology, 8*(10), 339–344.

Lynd, L. R., (1990). Large-scale fuel ethanol from lignocellulose. *Applied Biochemistry and Biotechnology, 24*(1), 695–719.

Marglin, S. A., & Schor, J. B., (1991). *The Golden Age of Capitalism: Reinterpreting the Postwar Experience*. Oxford University Press.

McGovern, P. E., Glusker, D. L., Exner, L. J., & Voigt, M. M., (1996). Neolithic resinated wine. *Nature, 381*(6582), 480, 481.

McGovern, P. E., Zhang, J., Tang, J., Zhang, Z., Hall, G. R., Moreau, R. A., Nuñez, A., et al., (2004). Fermented beverages of pre-and proto-historic China. *Proceedings of the National Academy of Sciences, 101*(51), 17593–17598.

Moyer, A. J., & Coghill, R. D., (1946). Penicillin: IX. The laboratory scale production of penicillin in submerged cultures by penicillium notatum Westling (NRRL 832) 1. *Journal of Bacteriology, 51*(1), 79.

Neushul, P., (1993). Science, government, and the mass production of penicillin. *Journal of the History of Medicine and Allied Sciences, 48*(4), 371–395.

Omemu, A. M., Akpan, I., Bankole, M. O., & Teniola, O. D., (2005). Hydrolysis of raw tuber starches by amylase of *Aspergillus Niger* AM07 isolated from the soil. *African Journal of Biotechnology, 4*(1), 19–25.

Pasteur, L., & Chamberland, R., (2002). Summary report of the experiments conducted at Pouilly-le-Fort, near Melun, on the anthrax vaccination, 1881. *The Yale Journal of Biology and Medicine, 75*(1), 59.

Pasteur, L., (1857). Mémoire sur la fermentation alcoolique. *Comptes Rendus de l Academie des Sciences, 45*, 1032–1036.

Plotkin, S., (2014). History of vaccination. *Proceedings of the National Academy of Sciences, 111*(34), 12283–12287.

Raper, K. B., (1946). The development of improved penicillin-producing molds. *Annals of the New York Academy of Sciences, 48*(2), 41–56.

Roberts, J. M., (1952). Antibiotic substances produced by species of *Cephalosporium*, with a description of a new species. *Mycologia, 44*(3), 292–306.

Robinson, D., & Medlock, N., (2005). Diamond v. Chakrabarty: A retrospective on 25 years of biotech patents. *Intellectual Property & Technology Law Journal, 17*(10), 12–15.

Samuel, D., (1996). Investigation of ancient Egyptian baking and brewing methods by correlative microscopy. *Science, 273*(5274), 488–490.

Shiio, I., Ôtsuka, S. I., & Takahashi, M., (1962). Effect of biotin on the bacterial formation of glutamic acid: I. Glutamate formation and cellular permeability of amino acids. *The Journal of Biochemistry, 51*(1), 56–62.

Sicard, D., & Legras, J. L., (2011). Bread, beer, and wine: Yeast domestication in the *Saccharomyces* sensu stricto complex. *Comptes Rendus Biologies, 334*(3), 229–236.

Smith, H. O., & Nathans, D., (1973). Nomenclature for restriction enzymes. *Journal of Molecular Biology, 81*, 419–423.

Soetaert, W., & Vandamme, E., (2006). The impact of industrial biotechnology. *Biotechnology Journal: Healthcare Nutrition Technology, 1*(7, 8), 756–769.

Susman, M., (1970). General bacterial genetics. *Annual Review of Genetics, 4*(1), 135–176.

Valamoti, S. M., Mangafa, M., Koukouli-Chrysanthaki, C., & Malamidou, D., (2007). Grape-pressings from northern Greece: The earliest wine in the Aegean? *Antiquity, 81*(311), 54–61.

Walsh, G., (2012). New biopharmaceuticals. *BioPharm International, 25*(6), 34–38.

Wang, Z., Zhuge, J., Fang, H., & Prior, B. A., (2001). Glycerol production by microbial fermentation: A review. *Biotechnology Advances, 19*(3), 201–223.

Zinder, N. D., & Lederberg, J., (1952). Genetic exchange in *Salmonella*. *Journal of Bacteriology, 64*(5), 679.

CHAPTER 3

Biology of Industrial Food Microbiology

3.1 INDICATOR MICROORGANISMS

A wide variety of diseases can occur by eating contaminated food with pathogenic microorganisms or their by-products; by no means can all of these diseases be classified as food poisoning. Additional definitions are currently in use in different countries within the UK and Europe; even the 1984 Food Act defines the term differently in various provisions. Two basic principles have guided us: firstly, the notion of 'poisoning' itself, and secondly, the distinction between food poisoning on the one hand and foodborne and food-related diseases on the other. The Oxford English Dictionary defines a poison as 'a substance which destroys life by fast action and when taken in a small quantity. Only a tiny proportion of the conditions we are concerned with are life-threatening in regular patients. Still, almost all are characterized by 'rapid action' – that is, the rapid onset of symptoms (typically nausea, vomiting, abdominal pain, and diarrhea) after ingesting contaminated food. As our title indicates, we deal principally with 'poisons' of microbial origin; to give a broad view of the general subject, we include agents such as viruses, mycotoxin fungi, and protozoa responsible for similar conditions to those caused by bacteria when present in food. We have specifically omitted organisms such as Salmonella typhi and the hepatitis A virus that produce characteristic diseases usually have long incubation periods. In the case of such named diseases, patients generally present with distinctive clinical features rather than gastrointestinal symptoms.

The majority of the organisms we consider food poisoning agents are also capable of rapid multiplication and or manufacturing toxins in many different foodstuffs (the exceptions being viruses and protozoa). In this respect, they can be distinguished from the causes of diseases such as brucellosis, where food is only the vehicle of infection and cannot support growth. We do not

Handbook of Industrial Food Microbiology. Manju Nehra & Vikash Nain
© 2025 Apple Academic Press, Inc. Co-published with CRC Press (Taylor & Francis)

deal with the majority of these foodborne diseases. Nor do we dwell on food-related diseases, which can be defined as certain chronic conditions of an immunological – as opposed to a gastroenteric-nature, caused either directly or indirectly by foodborne pathogens. In such situations, microorganisms in food can act as 'environmental triggers.' Auto-immune disease caused by tissue damage in organ systems may follow the organism's interaction or its products with the host immune system. Alternatively, immunological deficits or auto-immune disorders may be due to disrupted nutrient transport caused by an organism or its products interfering with intestinal integrity. Food-related conditions include septic and aseptic arthritis, Graves' disease, Guillain-Barre syndrome, myocarditis, IgA glomerulonephritis, and hemolytic uremic syndrome (HUS). Finally, mention must be made of bovine spongiform encephalopathy (BSE), a cattle disease similar to scrapie in sheep, probably caused by a so-called 'slow virus.' Despite its topicality and concern over the possibility of transmission from cows to humans via infected meat or milk, it has not yet been identified as man's disease. It appears in cattle to have far too long an incubation period (approximately five years) to conform to our notion of 'rapid action' as a significant characteristic of food poisoning. By our definition, therefore, microbial food poisoning is an acute condition, usually presenting as gastroenteritis; the first symptoms commonly arise within a few hours or a few days of consumption of food containing pathogenic microorganisms and or their products.

In most cases (the exceptions being viral and protozoal infections), the food in question will have supported the rapid growth of pathogens. The mycotoxin fungi can grow on various foodstuffs and be responsible for acute gastrointestinal disease. They are also associated with longer-term poisoning, their toxins giving rise to a range of chronic conditions, including cancer.

3.2 FOOD POISONING

Food poisoning is of many types and may be of microbial and non-microbial origin. The vast majority of reported cases of food poisoning are bacterial; in the UK, the four most common causes are Salmonella spp., Clostridium perjringens, Staphylococcus aureus, and Bacillus spp. (mostly B. cereus). Campylobacter spp. is also responsible for large numbers of food poisoning cases, although these organisms are not strictly reported as specific agents of food poisoning under present UK regulations. Most Campylobacter infections are sporadic from food sources and are rarely seen in large, publicized outbreaks. However, it should be stressed that they are, without a doubt, the most frequently reported bacterial cause of diarrhea in the UK.

When we try to understand the mechanisms of bacterial pathogenesis, it becomes apparent that food-poisoning bacteria can be split into two main groups: those whose modus operandi includes little evidence of toxin production and those that cause disease primarily by the production of toxins.

Over the last few years, there have been at least 50 general community food poisoning outbreaks of unknown etiology in England and Wales each year. It has been suggested that viruses may have played a role in some of these outbreaks. Viral food poisoning is still a very imprecise area. Cryptosporidium parvum's pathogenicity and role in opportunist infections have been highlighted in recent years, primarily due to the AIDS epidemic. These organisms have been responsible for a small number of food poisoning outbreaks and should not be overlooked when the more common microbial agents have been excluded. Certain food products, particularly those of tropical origin, may become contaminated with fungi, which then multiply and produce mycotoxins, some of which can be fatal, even in small quantities. Mycotoxins may be implicated in various diseases: recently, there has been considerable concern over the possible carcinogenic properties of aflatoxin associated with nuts and figs.

The causative agent of food poisoning may often be challenging to identify. Conclusive proof of a pathogen should require tests that comply with Koch's postulates (a set of conditions which should be fulfilled to establish that a given organism is the causal agent of a particular disease), but this is not always possible. Studies may be restricted to animal models and detecting in vitro invasiveness of virulence-associated genes. It may be that some bacteria are facultative pathogens, pathogenicity being dependent on such factors as the virulence of that specific strain, the infective dose, and the susceptibility of the patient. Bacteria that have been commonly incriminated and may be included under this heading are members of the Enterobacteriaceae (e.g., Klebsiella spp., Enterobacter spp., Citrobacter spp.) and Pseudomonas spp. Because of their uncertain significance, we do not consider these organisms in detail.

Those organisms which are currently of most significant importance, except for foodborne viruses, are primarily associated with foods of animal origin, viz. meats or dairy products, which become contaminated directly during milking or post-slaughter processing from the feces of infected carrier animals (Salmonella spp., Campylobacter spp.), or indirectly from the environment during food production (Clostridium botulinum, Listeria monocytogenes). In contrast, the transmission of E. coli and Shigella spp., etc., which are primarily of human origin, has become rare in most developed countries via contaminated food. However, indirect transmission of enteropathogens of human origin via sewage-contaminated shellfish is not unknown as a cause of outbreaks of viral gastroenteritis. When they occur, episodes of food poisoning may present in the form of:

- Sporadic cases with no apparent association with any other circumstances;
- Family outbreaks affecting two or more persons within a household; or
- Widespread community outbreaks affecting groups of persons who have eaten in the same restaurants or canteens, within residential institutions (hospitals, old persons' hornes, prisons, etc.), or who have eaten contaminated foods obtained from the same retail (or wholesale) outlet(s).

Diarrhea, vomiting, and abdominal pain are probably the most common symptoms of food poisoning. While other clinical features such as fever may be seen regularly, these three symptoms are significant.

Diarrhea may be defined as the malabsorption of salt and water resulting in a net 1,055 of fluid from the body (fluid secretion > fluid absorption) or the frequent evacuation of watery (or bloody) stools. It may occur when the intestinal mucosa is stimulated to secrete salt and water, when mucosal permeability is altered, following gross destruction of mucosal cells, or gastrointestinal motility disturbances and transit are disturbed. Infective agents may produce diarrhea by one or more of these mechanisms. Several bacterial toxins, e.g., cholera toxin, stimulate intestinal secretion and intestinal motility by altering intracellular levels of secondary messengers, e.g., cAMP. These increased levels of cAMP change the balance of Na+ and Cl– ions entering and leaving the enterocytes, which directly affect water absorption and secretion.

Vomiting is the forceful expulsion of gastric and intestinal contents through the mouth. Vomiting is usually preceded by nausea, excessive salivation, pallor, and sweating. The stomach divides into two sections in the vomiting process, and a deep inspiration is taken. The stomach contents are regurgitated up the esophagus with a muscular contraction of the diaphragm and abdominal muscles.

The coordination of the different physiological events during vomiting is controlled by the vomiting center situated in the brain's fourth ventricle.

The vomiting center can be activated by afferent impulses arising from many body parts, especially the digestive tract. Abdominal vagal afferent fibers can be subdivided into those originating from the gastrointestinal mucosa (the point of contact with the food and any toxins it may contain) and those that constitute tension receptors within the muscular layers of the gut wall. These mechanoreceptors signal the tension levels generated and contractile activity in the gut.

Unfortunately, little is known of the precise mechanism of abdominal pain production. The abdominal cavity organs have a lower number of nerve endings per unit of tissue than the skin; this contributes to the poor degree of localization of pain sensation in the abdomen. Moreover, it has been shown that inflammation makes nerve endings more sensitive to mechanical and chemical stimuli. Any microbial infection which produces an inflammatory response may, therefore, make the host more susceptible to painful stimuli such as those which result from distention.

3.3 ASPECTS OF BACTERIAL DISEASE

Salmonella is the most commonly reported cause of food poisoning in the UK, the USA, and many other countries worldwide. Since salmonella may be found in the intestinal tracts of man and animals, it is easy to understand how widespread contamination of meats and vegetables can occur and how difficult it can be to control infection. In many regions of the world, S. enteritis has now surpassed S. typhimurium as the most common serotype, however the reasons are yet unknown. Campylobacter is among the most common cause of food poisoning in developed countries. However, there are no official data for the UK and the USA, where such infections do not have to be reported to surveillance centers. The same organism is also responsible for enteritis in children and adults, which may not be associated with food. On the other hand, campylobacter infections are prevalent in young children in developing countries but are generally followed by immunity. Therefore, it is unlikely that older children and adults will be susceptible to campylobacter food poisoning than their counterparts in developed countries.

In contrast to salmonella, which produces many general community outbreaks, almost all campylobacter food poisoning infections are sporadic cases – an epidemiological nightmare. There is a very dose relationship between the incidence of indigenous Vibrio parahaemolyticus infection and areas of warm coastal waters. Dietary practices such as eating raw fish explain why this organism is the most observed cause of food poisoning in Japan. It is also common in certain parts of the USA but is rarely encountered in the UK. A similar relationship between other newly discovered Vibrio spp. Food poisoning cases and warm coastal waters have also been observed. Food poisoning caused by E. coli is reasonably uncommon throughout the world, though the number of reported cases may not be a reliable indicator of its prevalence in certain areas. In developed countries, E. coli food poisoning is not usually seen because sanitation and hygiene standards are ordinarily high. However, it is not always the case in developing nations, where fecal contamination is more widespread. In these places, laboratories may have difficulties distinguishing dangerous E. coli strains from the various non-pathogenic E. coli strains found in feces. In the case of Y. enterocolitis food poisoning, under-reporting may be a factor; this microbe can take more time to isolate and identify. It frequently fails to grow in typical enteric culture media at 37°C. This type of food poisoning is infrequently reported worldwide, but it is known that different serotypes are implicated in other countries. The importance of this organism lies in the fact that it can multiply at 4°C, the recommended temperature of most refrigerators. Another organism capable of low temperature (4C) multiplication is L. monocytogenes.

Although listeriosis is not particularly common, there have been notable outbreaks in Western Europe and the USA. This organism is significant as it may cause septicemia and or meningitis, and it has a high mortality rate of approximately 30%. Problems arising from contamination with S. aureus will be seen worldwide wherever humans handle cooked foods, especially if those foods are incorrectly stored after operating. Since inadequate training or refrigeration facilities are often implicated, S. aureus food poisoning is likely more widespread in developing countries than developed countries. C. botulinum was thought to be less of a problem in most developed countries, although many recent incidents have cast some doubt on this. However, a great deal is known about its ecology and the conditions necessary for toxin production. Generally speaking, this knowledge allows us to kill or inhibit the organism's growth. Botulism is much more prevalent in those areas of the world, such as Alaska and China, where traditional fermented foods are still widely eaten, and ignorance results in a lack of appropriate preventative measures. Therefore, this disease is frequently recorded throughout the world wherever such practices occur. If all foodstuffs were consumed directly after cooking, there would be less potential for toxins, leading to food poisoning. Notoriously, B. cereus has been most often associated with cooked cereals and cereal products (especially rice) stored incorrectly.

Diet management has proved successful in controlling symptoms of patients suffering from irritable bowel syndrome (IBS). This condition, which may involve abdominal pain and diarrhea, is caused by intolerance to specific foods or food components; it is not related to their microbial content. Diet management could also reduce exposure to potentially dangerous animal foods and their products (incredibly when raw), as they are responsible for most food poisoning outbreaks. From this perspective, it appears that eating a vegetarian diet is advantageous; nonetheless, other natural foodborne toxicants commonly occur in plants, aside from apparent microbial food poisoning hazards (e.g., listeria). Perhaps the most familiar natural toxicant is associated with the toxicity of green potatoes containing high glycoalkaloid concentrations. That may be responsible for neurological symptoms, including apathy, restlessness, drowsiness, and visual disturbances. The particular scientific finding indicates that potato glycoalkaloids possess gut-permeabilizing activity, which may point to the possibility of chronic effects of dietary glycoalkaloids; these will be enhanced in individuals who regularly eat large amounts of potatoes with peel. Some other plants that contain toxicants include cabbage, sprouts, courgettes (zucchini), beans, and cassava.

One question which is not perhaps immediately apparent, and is rarely considered, is the economic cost of food poisoning. The primary sources of financial losses are microbial spoilage between harvest or capture and consumption and costs associated with food poisoning disease. The expenses of spoilage are poorly documented and rarely quantified, but they are considerable and probably taken into account in the product's price. It is almost impossible to arrive at an accurate estimate of the cost of food poisoning since we know that most cases will go unreported, except for significant outbreaks. Even then, different analyzes include different real and apparent changes in their breakdown. Expenses that would have to be taken into consideration include medical care, including hospital costs, laboratory checks. Food poisoning can result from consumption of contaminated food carrying pathogenic bacteria or their by-products, but not all of these illnesses are classified as food poisoning. Various definitions are currently in use in the United Kingdom and Europe; even the 1984 Food Act uses the term differently in different clauses. The concept of 'poisoning,' as well as the distinction between food poisoning, foodborne, and food-related disorders, guided us. A poison is a drug that causes death.

Furthermore, economic consequences to the company or establishment involved are often catastrophic. It is more than likely that the total annual losses worldwide due to food poisoning will exceed several 100 million pounds. HACCP is a particular way of managing food manufacture and handling and covers product design, process design, and operating practices. Hazard analysis involves identifying factors that may dramatically affect food safety. Once these factors have been identified, various critical control points can be introduced during processing to improve monitoring and prevent or reduce risk to consumers. HACCP is one of many significant new and encouraging developments outlined in this hook that should help make the food we produce safer to eat.

3.4 FACTORS AFFECTING GROWTH OF MICROORGANISMS

Man has also always taken attempts to limit foodborne infections and spoilage, such as not eating some putrefied foods because they can be harmful and developing new ways to preserve foods in an edible state. Most procedures used today to preserve foods were established as a result of experience, and empirical understanding of food preservation processes dates back millennia. Cattle, for example, were primarily domesticated in ancient Egypt to give milk, which was then processed into cheese according to a schedule akin to a current code of practice. The pork was cured, fish were cooked and eaten or stored salted or dried. Some 2000 years BC, several different breads were baked, beer was brewed, and several other wines were made. However, within the last 200 years, we have understood that many processes that cause deterioration of plant and animal tissues involve microorganisms. Although microorganisms were observed and described by Leeuwenhoek in 1683, not until 1837 did Pasteur first associate bacteria with food spoilage. The demonstration that diseases were transmitted via foods also came in the 19th century. Hence, for most of man's history, food spoilage and disease transmission have been dealt with in ignorance of the responsible agents.

When agricultural products are harvested or animals slaughtered, the numbers and types of microorganisms they carry, which comprise the primary contamination, vary from one commodity to another. This variation depends on the geographical region, production methods, and harvesting and slaughtering methods. Much effort has been put into characterizing this initial microbial flora, expecting that it would be possible to exclude undesirable microorganisms, mainly those later associated with spoilage of the product or with food poisoning that can be traced to that product. The conversion of plant and animal tissues into food must prevent natural contaminants from either decomposing the produce or rendering it dangerous to eat because of the growth of microorganisms able to cause illness either directly or via toxins produced during multiplication. The organisms that become important in food poisoning or the spoilage of stored agricultural products are not necessarily relevant to animal or plant growth.

Conversely, some organisms that affect growth may be of little significance to the food production process: many plant diseases cannot be transmitted to man and are, therefore, of consequence only to the grower. Some animal diseases, though, are transmissible to man via food products, and in some areas, preventive action by producers has had a significant effect on human health. Eradication of animal diseases such as brucellosis and bovine tuberculosis has revolutionized the pattern of human diseases traceable to meats in some countries. However, it is worth noting that the factor most likely to persuade animal producers to take preventive action is that sick animals may die or not gain weight as fast as healthy animals. Care with particular procedures may limit, or even reduce, primary contamination, but eventually, a point is reached where the costs of those preventive measures outweigh the benefits. However, even healthy animals may still carry low numbers of bacteria able to cause illness in man. Current methods of transporting and slaughtering animals fail to minimize the dispersal of those pathogens into the food chain. From the point of view of the food processor, the primary microbial contamination of animals and plants is rarely under such control that there is an absolute assurance of complete freedom from particular hazardous microorganisms. In some instances, intensive production has markedly worsened the microbiological conditions of the raw product. A small slaughterhouse can be entirely adequate for small animals but unsuited to large numbers. Nor does mechanization necessarily improve hygiene or the microbiological condition of foods. Modern mechanized slaughterhouses have not provided raw meat less heavily contaminated with bacteria, partly because the emphasis remains on high throughput rather than high hygiene standards.

3.5 LETHAL EFFECTS OF TEMPERATURE

The most effective and extensively used approach for killing germs is heat treatment. On the other hand, heat resistance is dependent on the type of microbe. Heat inactivation and damage mechanisms are complicated because heat induces various changes in physiological material, making it difficult to pinpoint "the event" that causes harm or death. Furthermore, variances exist across strains of the same species, between life forms, and between spores and vegetative cells. Microorganisms have a great heat tolerance when growing at their optimal pH. Food quality can be assessed in terms of its nutritional value; most vegetative cells are killed at the sterilization temperatures generally used in the food industry, over 110 C.

In some cases, microorganisms can be sublethally injured by heat, making them more sensitive to adverse recovery conditions, such as organic acids, low pH, and low water activity. However, under these conditions, some microorganisms activate stress response mechanisms. As a result, cross-resistance is often seen, leading, for instance, to more excellent heat resistance among the microbial population.

Heat can activate or deactivate bacterial spores, resulting in delay, damage, mutations, or permanent deactivation. The pathogen is inactivated or experiences some type of damage when the temperature is raised over the optimum for growth, such as failure to initiate germination or replicate essential molecules. The germination mechanism may be harmed by high temperatures, which damage the trigger or cortical lytic enzymes. This could indicate that heat affects the membrane, encouraging antibiotic surface action. Recent research has found that physical damage to DNA does not occur in spores, implying that the DNA is shielded from heat in spores and that the process of thermal inactivation is not due to damage to those molecules. Several factors influence the high thermal stability of bacterial endospores. The major influences on microorganisms can be classified into three categories:

- Those that impact bacteria during sporulation.
- Those that are the result of treatments given before, during, or after the therapy and the medium's nature.
- Some of those who impact microorganisms during survivors' recovery.

Salmonella, Listeria, Campylobacter, and E. coli are essential vegetative cells related to the safety of heat-treated foods. These microorganisms can contaminate various foods, including meat, milk, vegetables, fish, and eggs. They can be a severe problem if the products are not adequately heat-treated and stored. D-values are used to measure heat resistance to pasteurization temperatures, and they vary with species and strain. It may be defined as the time needed to reduce microbial concentration by one log10. Microorganisms are more resistant to dry heat than wet heat; however, moist heat should be given special care in food microbiology because of the nature of food. Microorganisms' resistance is also affected by their physiological condition, with cells in the stationary growth phase being more resistant than cells in the exponential phase. Microorganism thermal inactivation, nutrient absorption, functional and organoleptic qualities, and simplicity of preparation are all affected by other parameters such as low pH, aw modification, and the presence of organic acids. The heat resistance of spores is determined by factors impacting microorganisms during the sporulation process. The key extrinsic factor contributing to improved spore heat resistance is the presence of metal ions (Ca2, Mg2, or Mn2) in the sporulation medium. The spores' heat resistance reduces as the presence of those ions decreases. A decrease in particular fatty acids can increase spore heat resistance during sporulation, whereas a shortage of glucose in the sporulation medium can diminish it. The heat resistance of spores varies depending on whether or not certain amino acids are present. pH is one of the most critical parameters impacting microorganism heat resistance throughout the thermal process. Acidic pH values lower spore heat resistance during the heating process, although thermal resistance does not vary significantly at pH values close to neutrality. pH and temperature have a complex relationship; some investigations have found that when treatment temperature rises, pH decreases. Other elements that can affect spore heat resistance during the heating process include NaCl concentration, water activity, sugar, or the presence of a germinant like lysozyme, which can assist raise the apparent heat resistance sublethal heating spores before the thermal process. Thermal process calculation is a fundamental step in ensuring the safety of thermally processed foods during their shelf life and optimization. The calculation has two main components: (i) microorganism inactivation kinetics; and (ii) heat transfer mechanisms in food. The inactivation kinetics of microorganisms is determined using mathematical models that describe the rate of inactivation at a specific temperature and the dependence of the rate kinetic parameter on temperature. It has generally been accepted that when a homogeneous microbial suspension of microorganisms (spores or vegetative cells) is subjected to a constant temperature, microorganisms die at a steady rate. This means that, at a defined time, the exact percentages of live microorganisms will die in the next time unit. According to these ideas, microbial inactivation by heat (spores and vegetative cells) follows two laws: (a) survivor law; and (b) thermal destruction (TD) law.

There were two traditional ways to calculate the thermal process, the general method and the formula method first used by Ball in 1923 and improved in 1957 by Ball and Olson, known as Ball's formula method. The available practice is more versatile and can be applied to almost any thermal situation. However, this method is time-consuming, and it soon gave way to the more convenient Ball's formula method. Ball's formula is convenient and rapid for estimating the sterilizing values of the process. The precision of the calculation, on the other hand, is determined by how well the heat penetration parameters adhere to the model's constraints. Heat penetration curves and parameters can also be derived analytically, eliminating the need for experimental heat infiltration assays when the retort or container size, for example, is modified. Parametric methods are the models used to perform these calculations. The two-dimensional partial differential equation describes convective heat transfer in a finite cylinder numerically using finite differences. The advancement of heat treatment procedures prompted research into new models and tools for developing and validating thermal processes, not only in terms of safety, but also in terms of nutrient retention and sensory qualities. As a result, the primary goal of thermal process optimization is to improve product quality while reducing unwanted changes and expenses. The varied heat sensitivities of microbiological and input impedance degradation rates are used in optimization theory. If heat is the primary criterion, preservation procedures can be separated into two categories: heating items in their final containers or packages and heating before packaging. Pasteurization, blanching, drying/concentration, boiling, and sterilizing are all examples of moist heat applications. Pasteurization must be accompanied with packing that prevents contamination and chilling that slows and stops the growth of surviving bacteria during the heat treatment for safety and quality. Thermal food processing aims to kill existing bacteria and their spores that could cause decomposition, foodborne sickness, or poisons. Experiments demonstrate that the logarithm of D and temperature have a nearly linear relationship (Figure 3.1).

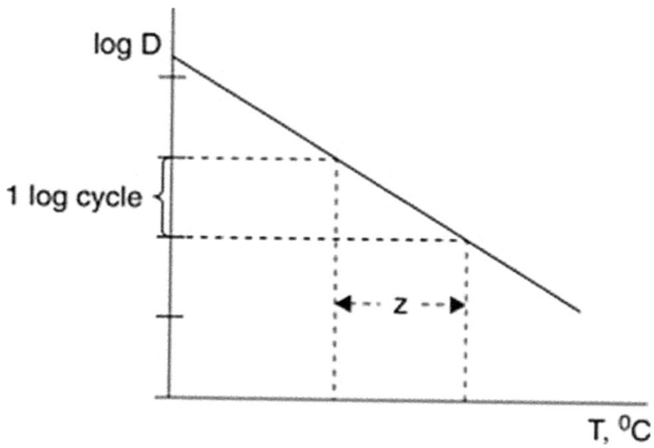

FIGURE 3.1 Effect of the temperature on D.

$$\log D1/D2 = T2 - T1/z$$

where; D1 and D2 denote the decimal reduction time at temperatures T1 and T2, respectively.

The temperature increment required for a ten-fold increase in the rate of thermal degradation is defined as the z value (i.e., for shortening D by a factor of 10). z=8–12°C is the optimal temperature for many spore-forming bacteria found in food preparation.

The Arrhenius model for the influence of temperature on the rate of chemical processes, written as the following equation, agrees with the logarithmic relationship between the rate of thermal inactivation and temperature:

$$\text{Log } k1/k2 = -E/2.3R(T2 - T1/T1T2)$$

where; k1, k2 are the rate constants at absolute temperatures T1 and T2, respectively; 'E' is the energy of activation; 'R' is the universal gas constant=8.31 J mol–1 K–1.

The activation energy E denotes the reaction rate's sensitivity to temperature, just as the z value denotes the rate of thermal inactivation's sensitivity to temperature. The following is the quantitative relationship between E and z:

$$E = 2.3RT1T2/z$$

If the absolute temperatures are not too far apart, then:

$$E \cong 2.3RTm2/z$$

where; Tm is denoted as average absolute temperature.

It's vital to remember that the 'energy of activation' for cell thermal death has no chemical meaning. The z value and the related energy of activation' of various thermal effects are documented with this warning in mind.

Representative approximate values of E, z for different thermal effects is listed in Table 3.1.

TABLE 3.1 Approx. Values of E and z for Different Thermal Effects

Thermal Effect	z (C)	E (10^3 kJ/kmol) at T_m = 393°K (120°C)
Cell death (spores)	9–11	300–330
Enzyme inactivation	14–20	150–200
Chemical reactions (not enzyme-catalyzed)	30–42	75–100

High economic losses and foodborne diseases may occur due to microbial contamination of food. Conventional thermal inactivation processes may not apply to fresh or raw food. However, these products are physiologically active food systems, and temperatures above 45°C can result in unwanted deterioration (e.g., fresh-cut salads). Because heat sensitivity of fresh food limits thermal inactivation processes and chemical treatments resulting in lower consumer acceptance, emerging inactivation technologies must be established to fulfill food safety requirements without affecting product quality.

The use of cold plasma as a method of inactivating foodborne microorganisms is an option. Considering proven plasma technologies, a rapid scale-up from lab to industrial appears feasible. Plasma switches used for power networks, high-definition large-area flat-panel displays, improved printability of foils via plasma, cost-effective light sources, modern television, energy-efficient lamps, and plasma etching for microcontroller manufacturing are just a few examples of industrial-use plasma technologies. Cold plasma technology is widely used in the non-food industry to decontaminate heat-sensitive objects such as electronics and medical devices and living vegetal or mammalian cells and tissues.

Alkaline phosphatase (ALP) is a phospho-monoesterase that catalyzes the hydrolysis of phosphoric acid monoesters (at an alkaline pH), resulting in phosphate salts' formation of the corresponding alcohol. This enzyme is widespread and can be found in various human body tissues (e.g., liver, kidney, bone, and blood cells). ALP is a membrane-bound glycoprotein containing a sugar moiety called sialic acid. ALP is frequently detected in milk and other bodily fluids from various organisms and animals. Because the concentration of this enzyme varies and its natural abundance and importance in biological systems, ALP assays are one of the most frequently performed tests for enzyme activity.

Purified bovine ALP has a molecular mass of 187 kDa and an isoelectric point ranging from pH 5.4 to 6.0. At 37°C, ALP has maximum activity in the pH range 9.65–10.1. The bovine ALP molecule forms zinc atoms that impart structural integrity and desirable functional properties in active form. The presence of metal ions contributes to bovine ALP's ability to reactivate following thermal inactivation. According to reports, bovine milk ALP is fat insoluble and is detected at the lipid–water interface in milk; ALP is connected with the milk fat globule membrane (MFGM) in raw milk. Although the breed of the cow, the volume of milk produced, and the time of lactation all affect the concentration of ALP in milk, researchers have established that the quantity of ALP in pooled milk is essentially consistent.

The dairy industry's reliance on bovine ALP is due to its thermal inactivation qualities. Due to its slightly greater resistance to thermal treatment than the bacterial pathogens used to test pasteurization procedures, thermal inactivation of milk ALP correlates well with thermally induced bacterial mortality. These connections established and substantiated the concept that the ALP assay might be used to verify pasteurization. Later studies demonstrated similar associations using comparisons of Salmonella senftenberg and Listeria monocytogenes inactivation responses. As a result, ALP activity is employed as a temporary, indirect measure of finished milk product pasteurization status. ALP activity in milk products can be detected and quantified using a variety of test procedures (colorimetric, fluorometric, chemiluminescent, and immunochemical). Only the colorimetric, fluorometric, and chemiluminescent assays have been confirmed as procedures for verifying pasteurization in the dairy business. Thermal inactivation of ALP follows a first-order rate expression, and the denaturation midway (half-life) of milk ALP is reached after 30 minutes of heating at 56°C. Additionally, indigenous milk ALP can be inactivated via high-pressure processing at a pressure of 400–800 MPa and a temperature of 5–40°C. The effect of pulsed electric field on milk can also be seen in the inactivation of ALP, both in its original form and in the equivalent ALP enzymes found in raw skim milk's natural microbial flora.

High temperatures profoundly affect the structural and physiological properties of sporulating and non-sporulating bacteria, with membranes, RNA, DNA, ribosomes, protein, and enzymes all affected. Nevertheless, no single event is responsible for cell death. Several factors contribute to the overall resistance to moist (wet) and dry heat; the latter, but not the former, induces mutations. Heat resistance develops during sporulation when spore-specific heat-shock proteins are also produced. Heat sensitivity is regained during the germination of spores.

Similarly, crops raised for local consumption present minimal spoilage problems. Still, if intended for a distant market, special care is often needed to protect them from further contamination and damage during transport. Efforts continue to minimize the occurrence of primary contamination with microorganisms able to cause illness, such as salmonellae, but attempting to control this initial flora is often less cost-effective than applying well-established preservative measures.

KEYWORDS

- **bacterial disease**
- **cell death**
- **chemical reactions**
- **enzyme inactivation**
- **food poisoning**
- **industrial food microbiology**
- **lethal effects**
- **microorganisms**

REFERENCES

Bigelow, W. D., (1921). The logarithmic nature of thermal death time curves. Journal of Infectious Diseases, 29, 528–536.

Boca Raton, FL: CRC Press.

Brul, S., Van, G. S., & Zwietering, M., (2007). Modeling Microorganisms in Foods. Cambridge/Boca Raton, FL: Woodhead Publishing Limited/CRC Press.

Gould, G. W., (1989). Mechanisms of Action of Food Preservation Procedures. London: Elsevier Applied Sciences.

Hallstro¨m, B., Skjo¨ldebrand, C., & Tra¨gardh, C., (1988). Heat Transfer and Food Products. London: Elsevier Applied Sciences.

http://pmp.errc.ars.usda.gov/PMPOnline.aspx (accessed on 18 November 2023).

http://www.campdenbri.co.uk/ (accessed on 18 November 2023).

http://www.combase.cc/index.php/es/ (accessed on 18 November 2023).

http://www.iftps.org/ (accessed on 18 November 2023).

https://irisk.foodrisk.org/Default.aspx (accessed on 18 November 2023).

Olson, J. C., & Nottingham, P. M., (1980). Temperature. In: Silliker, J. H., Elliot, R. P., Baird, P. A. C., Bryan, F. L., & Christian, J. H. B., (eds.), Microbial Ecology of Foods: Factors Affecting Life and Death of Microorganisms (Vol. I). New York: Academic Press.

Richardson, P., (2001). Thermal Technology in Food Processing. Boca Raton, FL/Cambridge: CRC Press/Woodhead Publishing Limited.

Richardson, P., (2004). Improving the Thermal Processing of Foods. Boca Raton, FL/ Cambridge: CRC Press/Woodhead Publishing Limited.

Smelt, J. P. P. M., & Brul, S., (2014). Thermal inactivation of microorganisms. Critical Reviews in Food Science and Nutrition, 54, 1371–1385.

Stumbo, C. R., (1973). Thermobacteriology in Food Processing. New York, London: Academic Press Inc.

Sun, D. W., (2012). Thermal Food Processing: New Technologies and Quality Issues (2nd edn.).

www.ift.com (accessed on 18 November 2023).

CHAPTER 4

Fermentation Systems

Fermentation is among one of the most ancient forms of energy metabolism. Fermentation refers to converting a complex molecule like carbohydrates into simpler compounds like organic acid or alcohol. The process of fermentation naturally occurs in various foods, and since ancient times, humans have deliberately utilized it to enhance the shelf life and sensory properties of food. Although, the term fermentation is also widely used in a larger context for the purposeful applications of microorganisms such as yeast, bacteria, and fungi to obtain products helpful to humans on a commercial level (e.g., enzymes, biomass, primary and secondary metabolites, recombinant products, etc.). Various methods have carried out fermentation depending upon the fermenter's size and the required product.

Bioreactors are the most important unit operation used in industrial bioproduction. Supplied material promotes the transformation of raw material into the desired product by microbial cell action or their cellular components like enzymes. A bioreactor is fundamentally a huge vessel, commonly made of thick stainless-steel body for the culture of microorganisms. Fermenter is used for the control system of the prokaryotic cells (bacterial) cultivation, while a bioreactor is employed to grow the eukaryotic cells (plant, mammalian, insect). Industrial fermenters are designed to regulate/maintain the conditions required for optimum growth and biosynthesis for industrially important microbial cultures. It also provides the ease of management for all operations carried out in the fermenters.

The bioreactors are constructed to provide the suitable environment for the development of products and their designed and operations are selected by scientists, bakers, or winemakers. Bioreactors are also known as heart of the fermentation system as all the product formation takes place in it and the condition required for efficient and proper microbial growth are regulated in this. It can be widely used for the various types of microbial products

Handbook of Industrial Food Microbiology. Manju Nehra & Vikash Nain
© 2025 Apple Academic Press, Inc. Co-published with CRC Press (Taylor & Francis)

such as for biomass production (such as Baker's yeast, single cell protein); for metabolite formation (like ethanol, pigments, organic acids, aromatic compounds and antibiotic); for substrates transforming or for production of an active molecule like enzymes. Depending upon the production scale and requirements, size of bioreactor can vary, and it can be as small as few mm^3 for microbial cell to large as 2–500 m^3 for plant scale fermenters. It may vary from 0.3–10 m^3 for pilot level fermenters and can be 100–1,000 ml shaking flask, and 1–50 liter for fermenters of laboratory scale. It regulates various conditions essential for fermentation including flow rates, gases level (air, oxygen, nitrogen, and carbon dioxide (CO_2)), pH, temperature, foam production, agitation speed or circulation rate, etc. Bioreactor has not only to maintain the favorable condition for biological activity but also have to minimize the undesired activities and it is the major challenge to maintain it.

Economic viability of a bioprocess and capital invested in it depend on the bioreactor's proper selection and design to ensure high process efficiencies.

A model fermenter should have an arrangement for controlling and regulating various unit operations like temperature, pH, adequate aeration, and agitation. Further, it should have low evaporation loss, low power consumption and proper facilities for sampling. Model fermenter should operate for a more extended period with minimum labor requirement. It should have a vessel of analogous geometry (small and larger vessels) that can be applied for varieties of reactions with appropriate product harvesting and cleaning process facilities. Modern fermenters are usually integrated with computers for efficient process monitoring, data acquisition, etc. (Figure 4.1).

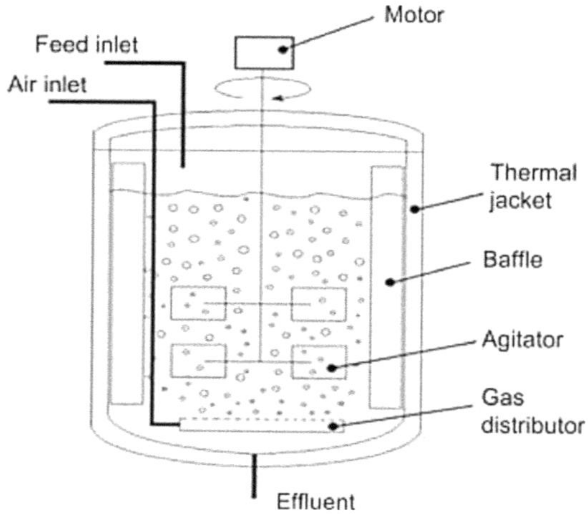

FIGURE 4.1 Typical diagram of a bioreactor.
Source: Reprinted with permission from Duan & Shi (2014). Copyright 2014. Elsevier.

4.1 ESSENTIAL FEATURES OF A BIOREACTOR

- The vessel strength should be enough strong to withstand the volumes of culture broth and pressure generated by gases produced during fermentation. It should be free from leakages to avoid contamination.
- Proper facilities for agitation and sufficient aeration of the fermenting broth should be provided for the optimum microbial metabolism.
- Should have an efficient arrangement for monitoring and controlling the temperature and pH.
- Have facilities to monitor and/or control dissolved oxygen in the reactants.
- It should have inner vertical plates (baffles) to inhibit vortex action. It should be coupled with inlet pipes to receive culture medium and inoculum of the microorganism.
- It should have facility for sampling and require a minimum of labor in maintenance, cleaning, operating, and harvesting operations.
- It should be constructed using the cheapest materials and adequate service provisions for individual plants.
- The size and the internal volume of the bioreactor are designed by bioengineers with proper knowledge of fluid dynamics, and the microbiologists should maintain the sterility of the entire system.

4.2 TYPE OF BIOREACTORS

Intensive understanding of the biological system (cell metabolism and growth, genetic manipulation and expression) and external factors affecting the natural system (physical and chemical environment) are highly required to design an appropriate bioreactor. Following the advancements in the understanding of biological systems, different types and configurations of bioreactor have thus been exploited and developed. The design of a bioreactor should be in a manner that it provides adequate sterilization conditions from in built within it. It also has adequate agitation, aeration, and product harvesting facilities with efficient drainage system. Fermenters/bioreactors are mainly divided into four groups: operational mechanism, microbial growth, process requirement, oxygen, and mixing requirements.

4.2.1 BUBBLE COLUMN BIOREACTOR

The bubble column reactor (BCR) is the simplest reactors as it is easy to assemble. BCR are pneumatic mixed reactors, specially designed for sensitive cells culture such as filamentous fungus cells. It is a cylindrical vessel with a gas distributor at the bottom having an aspect ratio of 4–6. In this type of bioreactor, gas is allowed to mix the content in the form of bubbles entering from the bottom side sparger in either a liquid phase or a liquid–solid suspension. These reactors are also known as slurry BCRs for a solid phase. These reactors find wide application in multiple reactors/contactors in various industries such as metallurgical, biochemical, chemical, and petrochemical industries.

These columns have low shear stresses, good mixing ability, easy operational process, and simple mechanical operations. Due to above-mentioned properties; these are best suited for the plant cell cultures. These BCRs are utilized in the industrial production of valuable enzymes, proteins, and antibiotic products. These reactors are finding applications in methanol production and manufacturing of other synthetic fuels (Figure 4.2).

- **Advantages:** Their advantages are high heat and mass transfer rates, making them more energy efficient. Their investment costs are low and are easy to construct. Further, these reactors have low shear forces due to absence of moving parts. BCRs are preferred for their high catalyst durability or other packing material. Online catalyst addition and withdrawal ability make them suitable for diversified applications.

- **Disadvantage:** The only disadvantage associated with BCRs is undefined flow patterns. It is difficult to control the flow rate. Experienced person is required for to maintain the optimum flow rate.

4.2.2 AIR LIFT BIOREACTOR

Airlift bioreactor mixes the content by using the expansion of compressed gas. They are also known as a tower reactor. Its name "airlift" is associated with the typical fluid circulation cyclic pattern created by pneumatic contact of the gas-liquid or gas-liquid-solid. It has two main types: external loop reactor and internal loop reactor. As the name indicates, the fluid circulation

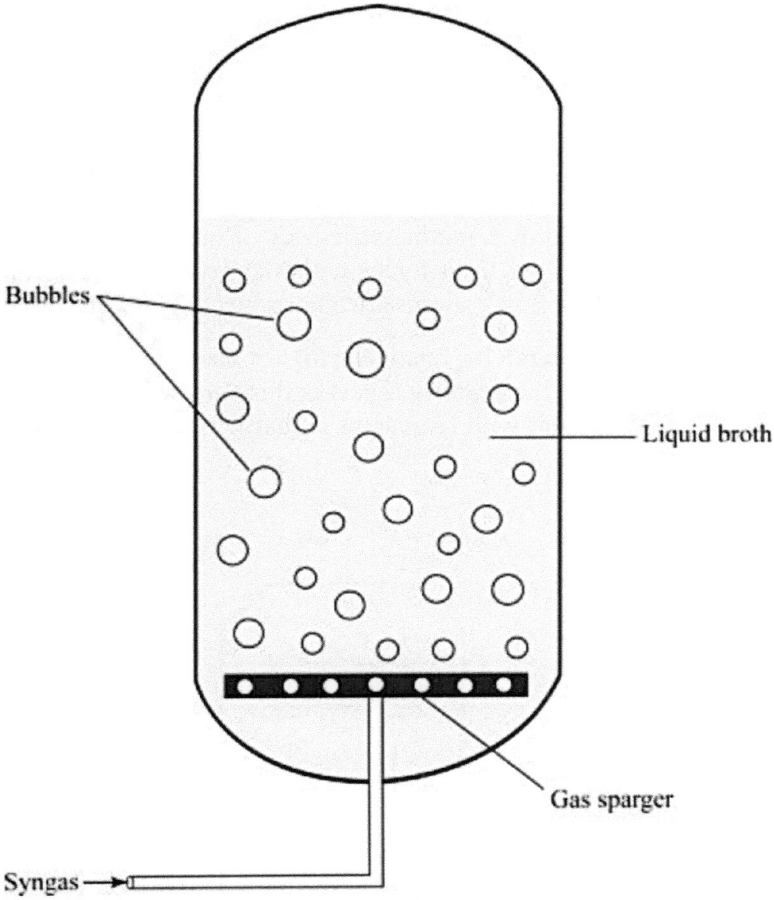

FIGURE 4.2 Representation of bubble column bioreactor.

Source: Reprinted with permission from Monir et al. (2020). Copyright 2020. Elsevier

is carried out in the outer loop reactor through distinct and separate channels. On the other hand, in an internal loop reactor, it is not separated in such a way. Here, a barrier is situated in a unic vessel that produces circulation channels or concentric tubes that create a central and peripheral channel.

Air-lift bioreactors of different shapes were designed for applications in several fermentation processes, single-cell protein (SCP) synthesis, plant and animal cell cultures, cell cultures, and wastewater treatment processes. Air-lift bioreactors are generally used in filamentous fungal fermentations because they have a moderate-shear environment that promotes mass transfer. The bioreactor airlift is the second category which is well defined and described, but less than that of stirred tank bioreactor. Numerous studies have investigated air lift-bioreactors with multiple draft tubes for their properties such as solid circulation, hydrodynamics, mixing and oxygen transfer rate, etc. (Figure 4.3).

- **Advantage:** The best advantage of this type of bioreactor is that only compressed air (or other gas) is required for the mixing purpose. Further, mixing efficiency of bioreactor make it more desirable. Low shear forces with high oxygen transfer rate and predictable flow patterns are other advantages of this bioreactor.
- **Disadvantage:** Its relatively higher cost due to compressing flow rate is the negative aspect of this bioreactor. Second associated problem with bioreactor is that it is difficult to adjust the flow rate.

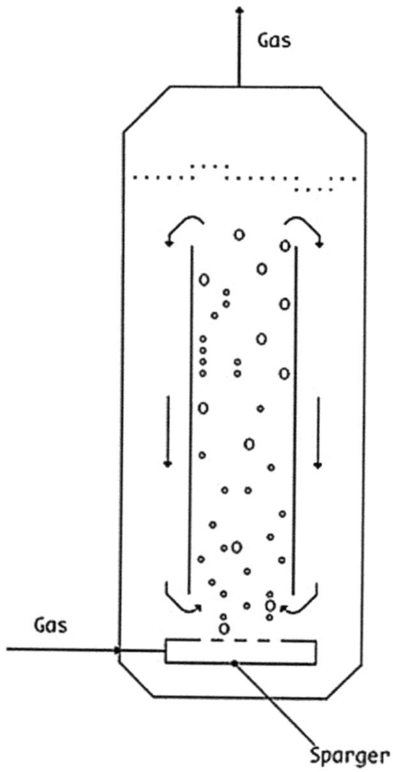

FIGURE 4.3 Principle of an airlift bioreactor.

Source: Reprinted with permission from Barragán et al. (2016). Copyright 2016. Elsevier

4.2.3 FLUIDIZED BED BIOREACTOR

Another commonly used bioreactor for the immobilization system is the fluidized bed reactor. This bioreactor substrate is supplied through a granular or pelleted biomass or enzymes suspended in encapsulated form in solid are mixed at sufficient speed by the help of gaseous phase inserted at the bottom of the reactor. The gas inlet is separated from liquid solid dispersion by a porous layer. In this system, when achieving the optimum conditions, its mixing efficiency is very much comparable to the agitated tank too, ,without the application of shear force. They are very large reactor ranging in size from 10–300 microns. To maintain the optimum fluid flow rate, it is necessary to design the bioreactor properly. A sufficient fluid flow rate is required to keep the catalyst particle suspended. In this type of reactor, catalyst is laid on the bottom part of the reactor. While reactants are supplied by a pump from the reactor through a distributing pump to remain bed fluidized. Bed behavior depends upon the state of the reactant. If reactant is in liquid state, its bed expands uniformly while expands in a non-uniform manner to make aggregative fluidization in the case of gas reactant.

In this system, the continuous processing between catalyst and reactant produces the product recovered simultaneously during the processing period. This type of bioreactors is primarily employed to utilize immobilized cells to process wastewater. Here, the cells are immobilized either by carrier or self-granulated, producing higher biomass retention with enhanced volumetric conversion efficiency. These bioreactors found their application in both the processes (aerobic and anaerobic) in wastewater treatment on industrial scale. The method is used for the treatment of both industrial effluent as well as domestic wastewater. Fluidized bed reactors are also widely used in micro-carrier cultures (Figure 4.4).

- **Advantages:** It has numerous advantages such as uniform gradient for temperature, easy catalyst replacement, and proper and homogenous mixing of reactant. Automatic and continuous operations in these bioreactors make them suitable for industrial purpose. Due to fluidization process, an optimum contact is maintained between catalyst and solid/gas reactant. Higher heat and mass transfer rate of these reactant make them much economic from energy point of view.

- **Disadvantages:** These bioreactors have some disadvantages as well. Some of them are listed here. They need high space for installation leading to higher capital investment. Second measure disadvantage is the requirement of pumping system to maintain the solid-liquid waste movement. For this purpose, a suitable pressure must be maintained throughout the processing period. Drop in pressure during processing is mainly faced problem with fluidized bed bioreactors. Erosion of internal components reduced its durability.

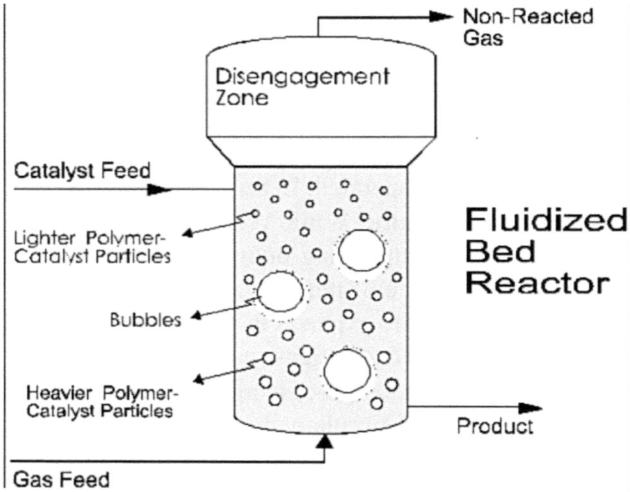

FIGURE 4.4 Fluidized bed bioreactor.

Source: Reprinted with permission from Fernandes & Lona (2000). https://creativecommons.org/licenses/by-sa/4.0/

4.2.4 MEMBRANE BIOREACTOR

Membrane bioreactors (MBR) have been in use since 1990. If defined broadly, a membrane bioreactor is a flow reactor in which cells and enzymes are separated through a membrane from the feed or product streams. The most essential characteristic of membrane reactors is retaining the enzymes/cells within it. This feature makes it suitable for continuous operation without the concern of enzymes/cells washing out from the reactor.

It is combination of ordinary system with the infusion of a filtration through a membrane. Addition of membrane results in the removal of suspended solid matters. Along with this, it also removes the nutrients from the reactant.

Membranes are made up of a wide range of materials, including cellulose, acetate, and nitrate, polyvinylidene difluoride, polysulfone, polypropylene, and polytetrafluoroethylene (PTFE), and polyacrylonitrile. Now a days, some other materials are also employed for the production of membranes with high efficiency and durability such as ceramic, silicone rubber, and ion exchange membranes. Different filtration can be used in this bioreactor. Generally used membranes are microfiltration and ultrafiltration membrane. Microfiltration membranes have pore sizes between 0.1 and 0.5 μm and can be used to confine cells within a reactor while imposing little restriction on the passage of soluble nutrients and products. Ultrafiltration membranes typically have pore sizes between 20 and 1,000 Å and retain or exclude macromolecules. Membrane bioreactor technology is evolving promptly in every part of the world research and commercial applications. It is abundantly used in the wastewater treatment from different sources. MBR have a more conventional arrangement for activated sludge process. Coupled membranes in bioreactor retain the biomass in this process.

MBR act efficiently in the clarification and considerably disinfect the industrial effluent due to its adequate pore size below 0.1 μm. Additionally it also reduces the requirement of large tank by concentrating the biomass and hence has better efficiency in biological treatment of wastewater (Figure 4.5).

- **Advantage:** MBRs are capable of working with concentration of mixed liquor suspended solids. Sludge productions in these bioreactors are comparatively low. It also facilitates the higher removal proficiency of chemical and biological oxygen demand.
- **Disadvantage:** The problem associated with MBRs is related to the membrane's functioning. Fouling of membrane is the fundamental limitation in its comprehensive application. Commercial applications of MBRs in wastewater treatment need the cheaper cost of the membranes.

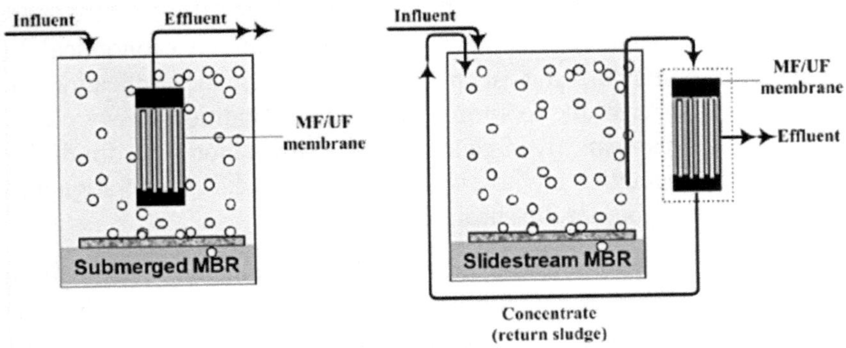

FIGURE 4.5 Membrane bioreactor.

Source: Alturki (2013). Published via open access.

4.2.5 STIRRER TANK BIOREACTOR

One of the most conventional bioreactors used is the stirred-tank bioreactor. Reactant and catalyst/microbial cell concentration are not associated with time in this fermenter. A fixed state can be attained either by turbidostatic or chemostatic principles. Flow rate modification is required to maintain the optimum substrate concentration, products, and microorganisms.

In the bioreactors, the homogenization and bubble dispersion are carried out by agitating the reactant mechanically. Such agitation needs a relatively higher energy consumption per unit of volume. The principal element of these bioreactors is the impeller or agitator, which executes a various function such as transfer of mass and heat, uniform mixing and aeration. Two types of agitators are commonly used in the fermentation industry, i.e., axial and radial flow impellers. These impellers generate different flow patterns in the bioreactors. Besides types of impellers, various other geometric conditions are essential for the optimum performance of stirred tank bioreactors. These include width and types of baffles, type, and position of the sparger, impeller off-bottom clearance, tank diameter to liquid height ratio, etc. Usually, 70–80% of the tank volume is occupied by liquid. If left as headspace for the removal of gases and foams produced during the processing Foams may be managed by applying foam breaker. Such bioreactors are used for both free and immobilized cells/enzymes. The sensibility of the biocatalyzers must be studied and analyzed in each case (Figure 4.6).

- **Advantage:** This type of reactor can be employed for high viscosity substrate while other bioreactors have limitations for high viscous solutions. It can also be used for immobilized enzymes with relatively lower activity. Better temperature control can be achieved easily in such bioreactors. The construction of stirred tank bioreactor is comparatively economical. Low manufacturing cost promotes commercial application—low labor cost due to simple operational conditions. Easy cleaning mechanisms give freedom to use a single bioreactor for different fermentation processes. Low shear force during operation makes it suitable for the shear-sensitive cells.

High capacity, better performance, and easily maintaining homogeneous conditions are other advantages of stirred tank bioreactors over other bioreactors. Due to these features, currently, these tanks are widely used in the biopharmaceutical industry to produce monoclonal antibodies and other therapeutics compounds of biological origin. High-capacity stirred tank bioreactors

(10,000–20,000-liter capacity) are being used widely in pharmaceutical industries for large scale production of animal cell cultures (Marks, 2003).

➢ **Disadvantage:** Main advantage associated with stirred tank bioreactors is the decomposition of some metabolites due to physical stirring produced through impellers.

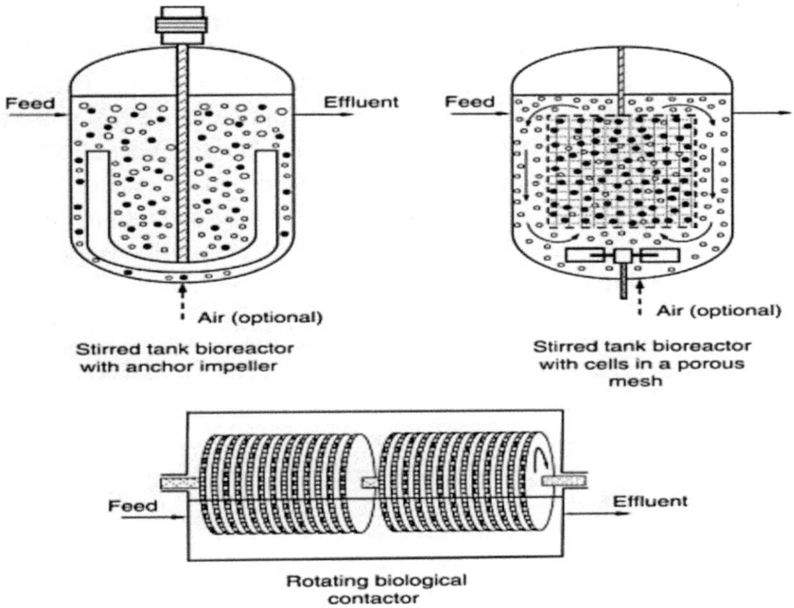

FIGURE 4.6 Stirrer tank bioreactor.

Source: Reprinted with permission from Nemati & Webb (2011). Copyright 2011. Elsevier

4.2.6 PACKED BED BIOREACTOR

Packed bed reactors, also known as fixed bed reactors, are widely used in chemical processing applications such as catalytic reactions, distillation, absorption, stripping, and separation. These bioreactors have a tube or channel-like assembly with the catalyst pallets or particle embedded inside. Liquid flows inside this channel co,mes in contact with the catalyst, and is converted into products. Catalyst/cells are immobilized by applying suitable carried and packed in the packed bed bioreactors. This arrangement results in high interfacial area for contact with solid-liquid reactants. Flow speed of the reactants over the static fixed particles is governed to provide sufficient time for reaction.

Packed bed bioreactors have significantly higher productivity within a compact size. These bioreactors are commercially utilized in gases and wastewater treatment and perfusion culture of immobilized mammalian cells. Many scientific studies demonstrated their potential application in artificial organ production in biomedical applications. Bio-artificial liver device production in a well-known example of such (Figure 4.7).

- **Advantage:** High reaction efficiency and simple operational procedure are main advantages of such bioreactor. Due to simple operational process, there is no need of much specified personal to operate it. Low construction and maintenance cost make them suitable for commercial application. Running cost effectiveness is other beneficial feature of packed bed bioreactor. Compact size of these bioreactors allows them to be installed at places with low area. Packed bed bioreactors can effectively work on even high pressure and temperature conditions.
- **Disadvantage:** These bioreactors' mass and heat transfer rate are comparatively low due to flow rate of liquid reactants. Accumulation of stagnant gas pockets produce gas flooding problem in the processing. This also results in poor liquid distribution in bioreactor. Gas pockets production limited its applications in aerobic fermentation.

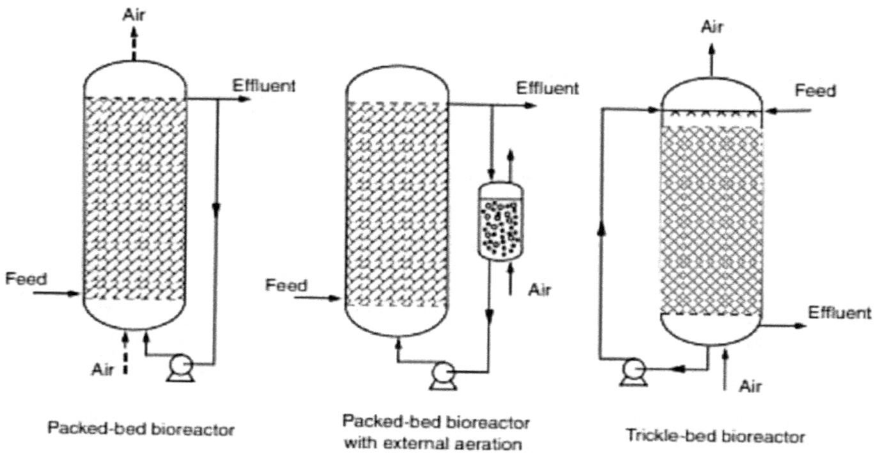

FIGURE 4.7 Packed bed bioreactor.
Source: Reprinted with permission from Nemati & Webb (2011). Copyright 2011. Elsevier

4.2.7 *PHOTO-BIOREACTOR*

Although various prototypes of photo bioreactors have been projected, only a few find their practical application in fungal biomass production. The central applications of photo bioreactors are in photosynthetic processes which involve vegetable biomass production algal production in controlled environment.

The algal biomass is used in aquaculture, treatment of wastewater, biosorption of heavy metals and CO_2 fixation production of fine chemicals and useful supplements in humans and animals. These bioreactors can also be employed to produce single cell protein from biomass.

The photo-bioreactor arrangements are used on both laboratory and industrial scale. Various models have been developed for the industrial as well as laboratory uses. Depending on their construction, these bioreactors can be classified as closed photo-bioreactor, open ponds, flat-plate, horizontal/serpentine tubular airlift, and inclined tubular photo-bioreactors. The main disadvantage is it has a limited application only for algal mass production. This requires proper processing conditions for mass algal cultures.

4.3 FERMENTATION METHOD

4.3.1 *SOLID STATE (SOLID SUBSTRATE) FERMENTATION*

Solid-state fermentation (SSF) is a microbial process in which fermentation raw material (substrate) is solid for growth of microorganisms. Besides, SSF was introduces for the development of alcoholic beverages and traditional food, now a days its application has been stretched to the biochemical and pharmaceutical sectors. Cereal grains, saw dust, bran, and other plants are commonly used substrates for the SSF process. The best-suited microorganisms for SSF are fungi and actinomycetes as they have large biomass and wider reach in substrate through hyphae. This technique also produces high-value products such as enzymes and toxins. Fungi are the source of most microbial enzymes production at cheaper cost through solid state fermentation. Recently, this method has been utilized to produce extracellular enzymes, certain valuable chemicals.

In this type of fermentation, the microbial distribution, growth, and development of products predominately occur on the solid surface. Based on chemical and physical features of the substrate, the moisture content may vary, but generally, it requires low moisture content. Heat produced from microbial growth and their metabolism increase the bed temperature, leading to moisture loss from substrate.

Proper forced/mechanical aeration is required to supply air and regulate temperature during fermentation. The bioreactors used in SSF are drum fermenters with continuous or staggered slow agitation, tray fermenters without any agitation, and column fermenters with forced aeration. The tray type fermenters may be wooden, metallic (aluminum, iron), or plastic trays with perforated bottom. These trays are sterilized and filled with a mixture of substrate and inoculum. These trays are arranged one above the other to an appropriate height. Optimum environmental conditions (temperature, humidity, etc.) are maintained throughout the fermentation. The column fermenter is made up of glass or plastic column having a water circulation jacket. Jacket is used to regulate the temperature. Such fermenters are aerated through forced air and are more expensive.

4.3.2 SUBMERGED FERMENTATION

Submerged fermentation methods can further be classified into various categories. Based on nature of fermentation process, there are mainly three types of reactions involved in fermentation process, i.e., batch, continuous and semi-continuous or fed batch depending on the feeding strategy of the culture and the medium into the bioreactor.

4.3.2.1 BATCH PROCESS

Batch stirred tank fermentation is the simplest type of fermentation. It represents a closed system where all the ingredients, i.e., medium, nutrients, and inoculums are supplemented in the bioreactor under aseptic conditions at the commencement of cultivation. In this type of fermentation, the volume of the culture broth is theoretically remained constant during cultivation in the bioreactor. Microbial cells undergo various phase of growth and multiplication during the period of incubation which further leads to compositional alterations in biomass, metabolites, and culture medium. In the initial stage, the microbial population remains contestant and grows after a certain inoculation period with a fresh medium. This stage is known as lag phase. When, the microbial cell numbers start doubling per unit period, it is assumed as log or exponential phase. A curve with a steady increasing slope is obtained when the cell number from that kind of reaction is plotted against time on a logarithmic scale. During this phase, the cells absorb most of the nutrients in the bioreactor. There is no net gain or decrease in cell number throughout the stationary phase. Energy metabolism and several biosynthetic activities continue in the cell. If the incubation is prolonged after the bacterial community has reached the stationary phase, the cells may die. Cell lysis, a much slower process than proliferation, can cause cells to perish. This type of fermentation is used for the high viscous substrate solutions and for the production of biomass where the cells need to be harvested at a particular growth stage. This type of fermentation is much suited for the use of immobilized enzymes which have low activity. It also finds applications in producing secondary or primary metabolites that require specific conditions for their optimum production. But main disadvantage of batch culture fermentation is its low production rate. It is only suitable for production of metabolites in small amount.

4.3.2.2 CONTINUOUS CULTIVATION

Continuous culture represents an open system where ingredients (nutrients and culture) are added continuously in bioreactor by maintaining the aseptic conditions. Fermentation products (biomass/metabolites) are also harvested simultaneously. Thus, the volume of the culture broth in the bioreactor remains almost constant due to constant rate of feed-in and feed-out. In this fermentation system, the growth rate of microbial cells is kept at a steady state for the long cultivation period. It is widely used for the large-scale production of metabolites. Although the microorganisms used in continuous fermentation process must be genetically stable, i.e., mutation free. It can be used for gas, liquid, or soluble solid substrates. Chemo stat is a common example of continuous process reactor.

4.3.2.3 FED-BATCH CULTIVATION

A combination of batch and continuous fermentation system is known as Fed-batch fermentation. It employs both modes of operations such as batch and continuous process. One or more required nutrients are supplemented in aseptic conditions in this fermentation system. It is a semi-open system where culture is supplemented step-by-step into the bioreactor. In this system, the volume of the liquid culture in the bioreactor increases with the time up to an optimum level. After a suitable time of fermentation (at the end of production cycle), the obtained product is collected.

Fed-batch systems overcome the problems associated with both continuous and batch systems. It can be employed particularly where substrate inhibition or catabolic repression is expected. The problem can be disabled by having a safe substrate concentration in batch mode initially. Then, it will be followed by adding the remaining substrate within the fed-batch operation. It can also be used in the case of the Crabtree effect (a high level of sugar concentration represses the yeast respiratory enzymes). It can be minimized by gradual feeding of the substrate. This will enable the elimination of ethanol produced by yeast in aerobic conditions. It is also suitable for a high cell density is desired. A high and constant specific growth rate can be maintained by the exponential feeding of the substrate into the system. Some other advantages of this system are high productivity, improved yield by controlled and progressive addition of nutrients, and continuous product synthesis.

4.4 ACID AND ALCOHOLIC FERMENTATION

Based on the involvement of oxygen, fermentation is of two types, i.e., aerobic and anaerobic. The final products of both the fermentations are of different chemical nature. Anaerobic fermentation produces alcoholic products, while acid-based products are obtained from the aerobic fermentation process. Due to their different products, they are also known as acid and alcohol fermentation.

4.4.1 ACID FERMENTATION

Acid fermentation can also be termed aerobic fermentation. In this type of fermentation, the reactants in oxygen are converted into organic acids. This fermentation method is used by animals and certain bacteria, like yogurt. This is also termed as fast fermentation. Due to involvement of sufficient amount of oxygen, reactants are entirely and quickly oxidized into final products, which are acidic. Aerobic fermentation, also termed aerobic glycolysis, is a metabolic process by which cells metabolize sugars by fermenting them in the presence of oxygen. This fermentation occurs through the repression of normal respiratory metabolism. Microbial production of organic acids and some amino acids used this type of fermentation. Presence of oxygen has a significant effect on the formation rate and yield of the product. Complete oxidation of reactants in aerobic formation results in the higher yield of product in comparison to the anaerobic or alcoholic.

There should be an appropriate provision of microbes-free air to be dissolved in a manner that can be available for the microbial growth. It should also have a flushing facility for gases (basically CO_2) produced during fermentation. The fermenters mostly employed in aerobic fermentations include stirred tank and air lift types. The fermenter should provide aseptic means for the withdrawal of culture samples during the fermentation and the introduction of inoculum at the initiation of the fermentation.

4.4.2 ALCOHOLIC FERMENTATION

Alcoholic fermentation is the most studied and known part of the fermentation process. It is involved in numerous necessary conversion, stabilization, and preservation processes for carbohydrate-rich substrates like fruit and fruit and vegetable juices. Alcoholic fermentation is propagated by yeasts and some other fungi and bacteria. In an anaerobic environment, only a fraction of the potential energy is produced because of incomplete oxidation. To obtain an equal amount of energy to that obtained in aerobic conditions, a high amount of glucose must be broken down under anaerobic conditions. This results in the high production of unoxidized organic byproducts during the fermentation process. Ethanol production is an example of alcoholic fermentation.

KEYWORDS

- **alcoholic fermentation**
- **bioreactor**
- **continuous cultivation**
- **fed batch cultivation**
- **fermentation**
- **membrane bioreactor**
- **photo-bioreactor**

REFERENCES

Abbott, M. S. R., Harvey, A. P., Perez, G. V., & Theodorou, M. K., (2013). Biological processing in oscillatory baffled reactors: Operation, advantages, and potential. *Interface Focus, 3*(1), 1–13.

Allen, J. W., & Bhatia, S. N., (2002). Improving the next generation of bioartificial liver devices. In: *Seminars in Cell & Developmental Biology* (Vol. 13, No. 6, pp. 447–454). Academic Press.

Alturki, A., (2013). *Removal of Trace Organic Contaminants by Integrated Membrane Processes for Indirect Potable Water Reuse Applications.* Thesis. https://ro.uow.edu.au/cgi/viewcontent.cgi?referer=&httpsredir=1&article=4759&context=theses

An, M., & Lo, K. V., (2001). Domestic wastewater treatment using immobilized sludge fluidized-bed reactors. *Journal of Environmental Science and Health, Part A, 36*(5), 819–831.

Ar, F. F., & Uysal, B. Z., (1998). Solid circulation in a liquid spout-fluid bed with multi draft tubes. *Journal of Chemical Technology & Biotechnology: International Research in Process, Environmental AND Clean Technology, 72*(2), 143–148.

Bakker, W. A. M., Van, C. H. J. L., Tramper, J., & De Gooijer, C. D., (1993). Hydrodynamics and mixing in a multiple air-lift loop reactor. *Biotechnology and Bioengineering, 42*(8), 994–1001.

Banerjee, S. (2023). Design of a Fermenter. https://microbenotes.com/design-of-a-fermenter/ (accessed on 18 November 2023).

Barragán, L. P., Figueroa, J. J. B., Durán, L. R., González, C. A., & Hennigs, C., (2016). Fermentative production methods. In: *Biotransformation of Agricultural Waste and By-Products* (pp. 189–217). Elsevier.

Bhattacharya, S. S., & Banerjee, R., (2008). Laccase mediated biodegradation of 2, 4-dichlorophenol using response surface methodology. *Chemosphere, 73*(1), 81–85.

Bio-Resource, (2012). http://technologyinscience.blogspot.in/2012/08/differenttypes-of-fermentors.html#.Wi0bMtKWbIU (accessed on 18 November 2023).

Botrous, A. E. F., Dahab, M. F., & Mihaltz, P., (2004). Nitrification of high-strength ammonium wastewater by a fluidized-bed reactor. *Water Science and Technology, 49*(5, 6), 65–71.

Chen, H. C., & Hu, Y. C., (2006). Bioreactors for tissue engineering. *Biotechnology Letters, 28*(18), 1415–1423.

Cinar, A., Parulekar, S. J., Undey, C., & Birol, G., (2003). *Batch Fermentation: Modeling: Monitoring, and Control*. CRC press.

Cosmol, (2015). https://www.comsol.com/blogs/packed-bed-reactor/ (accessed on 18 November 2023).

De Deken, R. H., (1966). The Crabtree effect: A regulatory system in yeast. *Microbiology, 44*(2), 149–156.

Doran, P. M., (1995). *Bioprocess Engineering Principles*. Elsevier.

Dürrschmid, M. P., Landauer, K., Simic, G., Klug, H., Keijzer, T., Trampler, F., Oudshoorn, A., et al., (2003). Comparison of fluidized bed and ultrasonic cell-retention systems for high cell density mammalian cell culture. *Biotechnology Progress, 19*(3), 1045–1048.

Faschian, R., De, S., & Pörtner, R., (2016). Multi-fixed-bed bioreactor system applied for bioprocess development of immobilized lactic acid bacteria. *The Open Biotechnology Journal, 10*(1).

Fernandes, F. A., & Lona, L. M., (2000). Fluidized bed reactor for polyethylene production. The influence of polyethylene prepolymerization. *Brazilian Journal of Chemical Engineering, 17*(2), 163–170.

Gonzalez, G., Herrera, G., Garcıa, M. T., & Pena, M., (2001). Biodegradation of phenolic industrial wastewater in a fluidized bed bioreactor with immobilized cells of *Pseudomonas putida. Bioresource Technology, 80*(2), 137–142.

Joshi, R., Sharma, V., & Kuila, A., (2018). Fermentation technology: Current status and future prospects. *Principles and Applications of Fermentation Technology,* 1–13.

Kantarci, N., Fahir, B., et al., (2008). Department of Chemical Engineering, Bogazic¸ University, 34342, Bebek-Istanbul, Turkey.

Kawasaki, H., Hirano, H., & Tanaka, H., (1994). Effects of multiple draft tubes with perforated plates on gas holdup and volumetric mass transfer coefficient in a bubble column. *Journal of Chemical Engineering of Japan, 27*(5), 669–670.

Kuila, A., & Sharma, V., (2018). *Principles and Applications of Fermentation Technology.* (pp. 3–13). John Wiley & Sons.

Kwon, Y. J., & Peng, C. A., (2002). Calcium-alginate gel bead cross-linked with gelatin as microcarrier for anchorage-dependent cell culture. *Biotechniques, 33*(1), 212–218.

Marks, D. M., (2003). Equipment design considerations for large scale cell culture. *Cytotechnology, 42*(1), 21–33.

Meng, F., Chae, S. R., Drews, A., Kraume, M., Shin, H. S., & Yang, F., (2009). Recent advances in membrane bioreactors (MBRs): Membrane fouling and membrane material. *Water Research, 43*(6), 1489–1512.

Monir, M. U., Yousuf, A., & Abd, A. A., (2020). Syngas fermentation to bioethanol. In: *Lignocellulosic Biomass to Liquid Biofuels* (pp. 195–216). Academic Press.

Nemati, M., & Webb, C., (2011). Immobilized cell bioreactors. *Comprehensive Biotechnology* (2nd edn., Vol. 2, pp. 331–346).

Nicolella, C., Van, L. M. C. M., & Heijnen, J. J., (2000). Wastewater treatment with particulate biofilm reactors. *Journal of Biotechnology, 80*(1), 1–33.

Pandey, A., Larroche, C., Soccol, C. R., & Dussap, C. G., (2008). Advances in fermentation technology. New Delhi: Asiatech Publishers, Inc. Chapter 7. *Bioreactors: Functions in Fermentation Processes* (pp. 172–201).

Pandey, A., Soccol, C. R., & Larroche, C., (2008). *Current Developments in Solid-State Fermentation.* Springer Science & Business Media.

Paulová, L., Patáková, P., & Brányik, T., (2013). Advanced fermentation processes. *Engineering Aspects of Food Biotechnology*, 89–105.

Resende, M. M., Ratusznei, S. M., Suazo, C. A. T., & Giordano, R. C., (2002). Simulating a ceramic membrane bioreactor for the production of penicillin: An example of the importance of consistent initialization for solving DAE systems. *Process Biochemistry, 37*(11), 1297–1305.

Santos, A., Ma, W., & Judd, S. J., (2011). Membrane bioreactors: Two decades of research and implementation. *Desalination, 273*(1), 148–154.

Şen, S., & Demirer, G. N., (2003). Anaerobic treatment of real textile wastewater with a fluidized bed reactor. *Water Research, 37*(8), 1868–1878.

Singh, J., Kaushik, N., & Biswas, S., (2014). Bioreactors–technology & design analysis. *The SciTech Journal, 1*, 28–36.

Singhal, G., Verma, V., Bhagyawant, S. S., & Srivastava, N., (2018). Fermentation technology prospecting on bioreactors/fermenters: Design and types. *Principles and Applications of Fermentation Technology,* 65.

Soccol, C. R., Pandey, A., & Larroche, C., (2013). *Fermentation Processes Engineering in the Food Industry*. CRC Press.

Spier, M. R., Vandenberghe, L. P. S., Medeiros, A. B. P., & Soccol, C. R., (2011). Application of different types of bioreactors in bioprocesses. *Bioreactors: Design, Properties, and Applications* (pp. 55–90). Nova Science Publishers Inc: New York.

Ugwu, C. U., Aoyagi, H., & Uchiyama, H., (2008). Photobioreactors for mass cultivation of algae. *Bioresource Technology, 99*(10), 4021–4028.

Veera, U. P., & Joshi, J. B., (1999). Measurement of gas hold-up profiles by gamma ray tomography: Effect of sparger design and height of dispersion in bubble columns. *Chemical Engineering Research and Design, 77*(4), 303–317.

Velizarov, S., Crespo, J. G., & Reis, M. A., (2002). Ion exchange membrane bioreactor for selective removal of nitrate from drinking water: Control of ion fluxes and process performance. *Biotechnology Progress, 18*(2), 296–302.

Visual Encyclopedia of Chemical Engineering, (2011). https://encyclopedia.che.engin.umich.edu/packed-bed-reactors/ (accessed on 18 November 2023).

Visual Encyclopedia of Chemical Engineering, (2013). https://encyclopedia.che.engin.umich.edu/fbr/ (accessed on 18 November 2023).

Wang, S. J., & Zhong, J. J., (2007). Bioreactor engineering. In: *Bioprocessing for Value-Added Products from Renewable Resources* (pp. 131–161). Elsevier.

Williams, J. A., (2002). Keys to bioreactor selections. *Chemical Engineering Progress, 98*(3), 34–41.

Xin, J. Y., Li, S. B., Xu, Y., Chui, J. R., & Xia, C. G., (2001). Dynamic enzymatic resolution of naproxen methyl ester in a membrane bioreactor. *Journal of Chemical Technology & Biotechnology: International Research in Process, Environmental & Clean Technology, 76*(6), 579–585.

Xu, F., (1999). Fermentation, biocatalysis and bioseparation. *The Encyclopedia of Bioprocessing Technology*, 1545–1554.

CHAPTER 5

Fermentation Raw Materials

5.1 MICROBIAL FERMENTATIONS

Yeast ferments barley malt into beer, and human muscles work during heavy exercise in the limited oxygen supply. Do you ever think how this happens in the absence/scarcity of oxygen?

Both of these processes may occur when normal, oxygen-dependent (aerobic) cellular respiration is prevented, i.e., when oxygen is not present to act as a receiver at the end of the electron transport chain (ETC). Glycolysis serves as the central process in these fermentation pathways, with additional reactions grafted on at the very end. Alcohol is produced through several reactions in yeast, whereas lactic acid (LA) is produced by similar events in your muscles.

While not the only method, fermentation is a well-liked method for obtaining energy from anaerobic process. Some biological processes use a mineral (inorganic molecule) other than oxygen as the last electron receiver for an electron transport system, such as sulfate. Some archaea and bacteria engage in what is referred to as anaerobic cellular respiration. Numerous species and cell types use fermentation, another anaerobic (non-oxygen-requiring) pathway for glucose breakdown. The only power conversion pathway throughout fermentation, with one or two extra processes patched on at the end, is glycolysis.

In every aerobic respiration route, oxygen, or O_2, serves as the final electron acceptor. If aerobic respiration continues, ATP will be created by the ETC using high-energy electrons (carrying by NADH or FADH2). In order for the glycogen synthesis process to proceed if aerobic respiration is absent, NADH must be reoxidized to NAD+ and employed once more as an electron carrier. The final electron acceptor in various biological processes

Handbook of Industrial Food Microbiology. Manju Nehra & Vikash Nain
© 2025 Apple Academic Press, Inc. Co-published with CRC Press (Taylor & Francis)

is an organic molecule. The process of converting an organic molecule into NAD+ from NADH is known as fermentation.

As a final electron acceptor, however, some living systems use an inorganic molecule. Because both processes involve organisms generating energy for use without oxygen, they are together known as anaerobic cellular respiration.

5.1.1 ANAEROBIC CELLULAR RESPIRATION

Some prokaryotes, including bacteria and archaea, use anaerobic cellular respiration. Both soil and domestic animals' intestinal systems contain these microorganisms. In a similar manner, Archaea and anaerobic sulfate-reducing bacteria convert sulfate to hydrogen sulfide to create NAD+ from NADH.

Many different fermentation techniques are used by bacteria. Some prokaryotic cells are facultatively anaerobic, which means they can alternate between aerobic cellular respiration and fermentation depending on the availability of oxygen. Prokaryotes like clostridium and others are required anaerobes. Obligate anaerobes can only live and develop in the absence of oxygen molecules because it is harmful to them. With the exception of LA fermentation, almost all other types of fermentation result in the production of gas. When certain carbohydrates are fermented, a particular type of gas is created that can be used as a biomarker and as a technique for assessing bacteria in the lab. Diverse species use a variety of fermentation mechanisms to ensure an adequate supply of NAD+ for the sixth step of glycolysis. Without these routes, that phase would not take place, and the breakdown of glucose would not result in the production of ATP.

Another typical process is alcohol fermentation, which produces alcohol from ethanol. The conversion of pyruvate to ethanol takes place in two steps. A carboxyl group is removed by pyruvate and released as carbon in the first phase (Figure 5.1).

Many types of bacteria are facultative anaerobes, meaning they may alternate between aerobic and anaerobic conditions depending on the availability of oxygen. They can use their glucose molecules to produce more ATP when oxygen is available since aerobic respiration generates more ATP than anaerobic respiration, but they can still metabolize and survive when oxygen is scarce.

Anaerobes are a necessary component of other types of bacteria and archaea. They don't need oxygen to live and develop. When exposed to

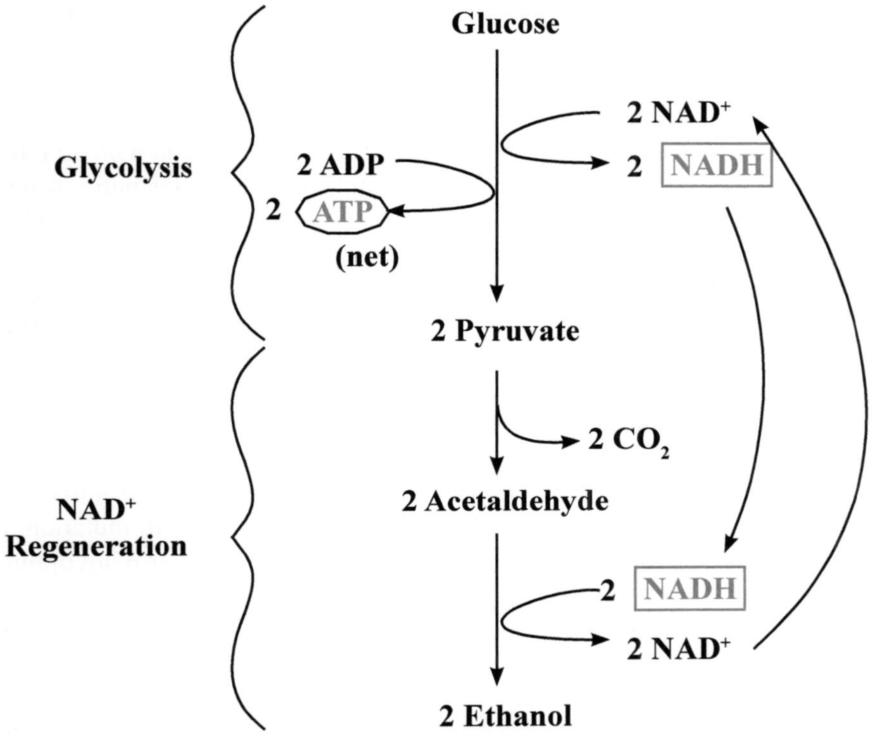

FIGURE 5.1 Carboxyl group removed by pyruvate.

oxygen, these microorganisms suffer damage or are eradicated. For instance, the Clostridium bacteria that result in food illness such as botulism are obligate anaerobes squared. Recent research has revealed a number of multicellular organisms lacking oxygen in deep-sea sediments.

5.2 SINGLE CELL PROTEINS

Yeast, fungi, algae, and bacteria are examples of dried, dead microorganisms that fall under the category of single-cell protein (SCP). In order to reach their optimum protein content, these bacteria exhibit their fastest development on a variety of carbon-rich sources. In 1967, Carol Wilson replaced the less appealing terms "petro protein" and "microbial protein" with the phrase "single-cell protein." The vast majority of bacteria utilized to extract SCP are unicellular.

SCP offers several benefits over flora and fauna proteins, including the fact that it can be generated all year round and that its development requirements are not influenced by the seasons or climate. SCP's use as a nutrient supplement for humans is advantageous for the environment because it can be grown on trash, has a good nutritional value (protein) with a broad range of amino acids, a low nutrient content (lipid), and a greater nutrient ratio (protein and lipid) than roughage. A high concentration of nucleic acids is also helpful. Their use as a global food is further constrained by the existence of an indigestible cell wall, unappealing colors and flavors, and a higher contamination of food and cell recovery. In consequence, efforts are being made to prevent risks that could increase public acceptance of SCP cultivation methods.

Proteins from single cells are often produced during fermentation. This technique for generating and creating cell clumps involves isolating and culturing particular microbe strains on suitable substrates. The initial step in creating a method is microbial screening. Using selection, mutation, or other genetic techniques, is the process of selecting and improving optimal production strains from the air, soil, water, or swabs of mineral (inorganic) or other material. As a result, all biochemical pathways and cell morphologies for the optimized strain as well as its technical culture parameters are determined.

Process engineering and equipment technologies also modify the process' technical performance to prepare the end product for widespread use. External factors (cost, energy) come into play. Environmental protection, process safety, and product safety issues are all taken into account while creating SCPs. Finally, among the legal and controlled components of innovation safety and security are operational licenses, product authorization for certain uses, and the legal protection of ongoing developments and microorganism strains. A type of oxygen-free fermentation is water fermentation. The submerged method in any way maintains the material for fermentation in a liquid condition, including the nutrients required for growth. The continuously operating substrate-involved fermentor removes the biomass regularly from the fermentation tank utilizing a variety of techniques. After purification or centrifugation, the product is dried. It is impossible to exaggerate the significance of oxygenation in the farming process. Before being dried, the substance is filtered or centrifuged. The cultivation process involves crucial aeration. Heat is produced during cultivation and is subsequently expelled using a cooling system. Several methods can be used to extract microbial biomass. Filtration is used to collect filamentous fungus, while centrifugation is used to recover single-cell organisms like yeast and bacteria. Retaining as much water as you can is essential before ultimate drying in sanitary settings.

5.2.1 SEMISOLID FERMENTATION

In solid-state fermentation (SSF), such as with cassava detritus, semisolid fermentation—which does not necessitate substrate removal—is more common. A submerged culture fermentation is more expensive to set up and maintain than a surface culture fermentation. In the culture process, heat is transferred from the liquid phase to the environment while a multiphase system is being stirred and mixed, oxygen is delivered red to the microorganisms through air pockets in the liquid phase, and so on. A special type of bioreactor called a U-loop fermentor was created to track mass and energy transfer occurrences. One of the fundamental steps in the production of SCP is the preparation of an appropriate medium with a workable source of carbon, the protection of cross-contamination of the medium, the selection of an appropriate fermentor, the choice of microorganisms with desirable traits, separation, and processing of the produced biomass. Alkenes, methanol, gaseous hydrocarbons, ethanol, renewable sources such as carbon oxide molasses, brewery effluents, polysaccharides, and other solid materials are examples of carbon sources.

SSF has been used to produce a variety of value-added products, including SCP, feeds, enzymes, ethanol, organic acids, B complex vitamins, pigments, and flavors. Several articles have reported various types of fermenter designs, operational parameters, and microorganisms (SSF). When a solid culture substrate has been seeded with microorganisms, it is often placed on flatbeds. The substrate is then kept in a temperature-controlled setting for a few days after that. Liquid state fermentation occurs on an industrial scale in tanks that are 1,000–2,500 square meters in size. Growing single-celled organisms like bacteria and yeast in liquid culture is appropriate. The bacterium must be continuously supplied with oxygen for liquid aerobic fermentation to occur, which is typically achieved by stirring the media. It is crucial to correctly control temperature, soluble oxygen, ionic strength, pH, and nutrition management in order to accurately direct the creation of desired metabolites. The preparation of the medium, fermentation, separation, and downstream processing are the three primary processes in this process.

5.2.2 APPLICATIONS OF SINGLE CELL PROTEINS

SCP is thought to give immediate energy. SCP usage is ideal for bright skin and healthy eyes. They are regarded as the greatest diet with added protein for undernourished kids and teenagers. They are also an excellent source of vitamins, minerals, amino acids, and other nutrients. SCP is used in natural and medicinal treatments to treat obesity and manage weight. They reduce stress, cholesterol, and blood sugar levels in diabetic people. In addition to being utilized in cuisine, they are also found in a variety of herbal cosmetic products including Biolipstics and herbal face cream. Another benefit of SCP technology, aside from its nutritional content, is that it can be made at any time of year and helps with waste management by using waste materials as a substrate. A tiny plot of land requires less time for development. SCP is a superfood because it is beneficial following a food crisis.

5.2.3 DISADVANTAGES OF SCP TECHNOLOGY

The SCP exhibits extraordinary properties as a food source for humans and animals, but a number of problems impede its widespread usage. There could be certain problems, such as the fact that all fast-growing organisms have higher nucleic acid concentrations than do traditional protein sources.

The issue with consuming a lot of nucleic acid-containing protein (17–25 g/100 g protein dry weight) is that it significantly raises blood uric acid levels and causes health issues including kidney stones and gout. These problems have also been connected to the shortcomings of SCP derived from bacteria, which include a high ribonucleic acid concentration, a sizable risk of contamination during the manufacturing process, and challenges in retrieving the cells. Also worth mentioning is the possibility that the microbial cell wall is inedible.

- Algal single cell protein, with the exception of Spirulina, has a high chlorophyll content and is therefore unfit for human consumption.
- It is poorly compacted, with only 1–2 g of dry weight per liter of substrate, and fermentation contamination is a major concern.

5.3 BIOFERMENTOR CONTROLS

Fermentor design and fermentation method both affect how well a fermentation system performs its intended tasks. In order to create a rational fermentor, it is important to carefully consider how the fermentor will fit into the overall process design. From the beginning, it is important to carefully consider the plant's running hours, space restrictions, relationship between bioreactor production and downstream equipment throughput rates, containment and validation requirements, utility requirements, potential plant breakdowns, manpower requirements, capital investment, and running costs. For production equipment, consistency, safety, affordability, and adherence to legal requirements are typically the top priorities. As the name suggests, a fermentor appears to be something that ferments in general. For thousands of years, people have been aware of fermentation.

However, it has mostly been used to produce alcohol by fermenting starch from various sources, including fruits, grains, and tubers. A fermentor, however, has only been described in recent years as the perfect environment for bacteria and/or fungi to flourish, with the cultivation of those organisms producing the desired output.

A bioreactor is a container used for chemical reactions involving living things or substances produced from them that are biochemically active. Bioreactors are frequently cylindrical, made of stainless steel, and range in size from a few liters to cubic meters. In a nutshell, bioreactors are large-scale operations with volumes/capacity that can reach several liters. Mammalian or insect cells are grown and maintained in a bioreactor. Fermentor, on the other hand, is a method for growing and maintaining bacterial or fungal cells. Taller vessels are employed for bacterial activities in bioreactors to optimize mass oxygen transfer, which is another geometrical distinction between them and fermentors. For mammalian cell culture, shorter containers improve mixing.

5.3.1 BIOFERMENTOR CONTROLS

The most popular manufacturing processes in the biotech industry are bioreactor or fermentation processes. Due to lost items and downtime, implementation delays and process disruptions can lose millions in sales. Due to the bioreactor's significance, a biotech company's capacity to remain profitable depends on getting it into production as soon as is practical and keeping it running.

Many end users make an effort to minimize the I/O footprint of their control devices during installation because bioreactors require a wide range of signals. Analog I/O points are used to measure buffer and media flow rates, pressure, and temperature. Pumps and valves used for metering and peristalsis are controlled by discrete I/O. Additionally, analytical probes control conductivity, dissolved oxygen, and pH. To minimize shear, bioreactors must maintain a precise control speed in the agitator throughout the process. The produced microbes could be shattered if the agitator generates too much turbulence. Poor agitator design and high oxygen flow rates through the sparge tube could make the shear issue worse.

5.3.2 TEMPERATURE CONTROL

The thermal management system is created using PID control algorithms. The liquid temperature is tracked and contrasted with the target temperature to provide the error signal. Control algorithms are used to process the false signal and deliver the desired output to the heater driver. The feedback control process operates with the thermal system to maintain the target temperature by reducing the differential to zero. In accordance with the error indication, continuously variable controllers power the heater. Energy supplied to the heater steadily decreases as the measured value gets closer to the desired value. PID controllers are the most popular type of temperature controller.

5.3.3 PH CONTROL

Bioreactors use on-off pH regulation in the real world. The reagent addition control element is set up in the two-position control system to be in one of two states at all times: completely open or completely closed. In managing pH, the retention time needed to permit the neutralization reaction is a crucial factor.

5.4 IMPORTANCE OF RAW MATERIAL IN FERMENTATIONS

The specialty market for fermentation products is dominated by the chemical and pharmaceutical sectors. The ancient "independence fermentation industry" is quite small; the few examples that survive today are identified among citric acid and enzyme makers, as well as biopharmaceutical companies, to the extent that the large global players have not yet acquired them. The scenario is different for the initiatives using LA and bioethanol. These products have attracted investment from the agro-industry, specifically from U.S. corn wet-millers, who are claiming this sector as their original area of diversification. The chemical sector, on the other hand, is more hesitant and insists on using their preferred raw materials and tried-and-true petrochemical processing methods. In any event, much R&D and conceptual work are required, but the potential for fermentation to contribute to the production of organic compounds is very great. The United States is once again in the lead when we take into consideration the Sustainable Biomass feedstock Project of the Energy Department, which, among many other things, supports the development of new bioreactors for a total of 12 alcohols and carboxylic acids. In my opinion, individuals that get started today could be able to maintain a somewhat comfortable firm in a rapidly changing economic, ecological, and administrative environment after the year 2000.

Theoretically, the best methods for transforming highly oxidized natural feedstocks like carbohydrates into specific chemical products are biochemistry and industrial biotechnology. This assertion is amply supported by the minimal results of organic synthesis in the conversion of renewable resources, with the exception of fats and oils. Though the quantity of biotechnological conversion items in the market for high-end, low-volume specialties is quite outstanding, biotechnology's influence on the overall industry is still quite little. Only a few large-scale items are produced using biotechnology, despite the fact that about 100 medicinal and nutritional specialties are produced by fermentation, microbial transformation, and enzymatic synthesis. These include glucose from enzymatic starch hydrolysis, bioethanol, citric acid, monosodium glutamate, and lysine. However, the bulk market is particularly important in terms of substituting crude oil, environmental effects, or as a substitute market for agricultural surpluses. Instead of a shortage of raw materials or customer needs, biotechnology's restricted impact on the transformation of carbohydrates into manufactured products can be attributed to a lack of goods, product ideas, and applicable manufacturing methods. Technology shortcomings must be taken into consideration because of this severe shortage of manufacturing viability and the resulting efficiencies in contrast to the well petrochemical products. Nevertheless, for those who are familiar with it, biotechnology is a widely used industrial tool. To assess how successfully carbohydrates and other agricultural raw materials may be transformed into existing or new chemical compounds, several crucial success variables can be extracted from existing procedures.

With the exception of a few instances, such as the conversion of glucose into gluconic acid, sorbitol into sorbose, and sucrose into dextrane or lavane, which actually more closely resembles microbial conversion than fermentation, the structure of the carbohydrate substrate of fermentation processes is no longer discernible in the final product. The C_6 moiety is typically transformed into primary metabolites, such as ethanol, acetic acid, citric acid, and amino acids, or secondary metabolites, such as antibiotics, alkaloids, or other heterocyclic compounds, through 10 to 20 enzymatic stages in a fermentation process (Figure 5.1). Similar to this, there are numerous enzymatic steps required to transform the C_6 moiety into microbial carbohydrates like xanthan or non-polysaccharide type polymers like poly-p-hydroxybutyrate.

Efficiency on substrate and output in terms of volume. The main source of carbon and energy for the organism during a typical fermentation is carbohydrate. Nitrogen is also obtained from proteins, amino acids, nitrate, ammonia, and other compounds. As a result, a certain quantity of carbohydrate substrate must be transformed into energy and carbon dioxide (CO_2), which unavoidably leads to a molar product yield per substrate below 1.0. The carbohydrate substrate serves primarily as an energy source in anaerobic fermentations. The carbohydrate is only partially oxidized in the absence of oxygen. As metabolic waste, fermentation end products like ethanol and LA are released in large amounts into the fermentation medium. To achieve the best resource management, carbohydrate consumption is closely controlled in aerobic bacteria. The synthesis and accumulation of products are in intense competition with the requirement for carbohydrates for growth and energy metabolism. Overcoming the genetic and metabolic regulation of the specific synthesis route is a requirement for metabolite and final product oversupply. The quantity of enzymatic steps required for the synthesis of carbohydrate feedstock and the amount of different metabolic outlets that compete for the carbon precursors supplied by the central metabolism further diminish product output.

Increasing product yield per substrate and process volume productivity is the most crucial step in creating industrial fermentation systems. By optimizing the process engineering overall and the microbial production strain, this must be accomplished in a coordinated manner. By using genetic engineering, traditional mutagenesis, and strain selection, the absorption of carbohydrates, the regulatory pattern, and the absolute activity/concentration of the required enzymes of the corresponding synthetic route may all be improved (Figure 5.2). Moreover, strain selection and mutation can reduce the production of undesirable byproducts. Similar to this, the overall fermentation conditions, including the choice of substrate, substrate concentration, substrate feeding, pH, and temperature profile, agitation, and aeration, have a significant impact on the productivity of the system overall as well as the yield of product on carbohydrate substrate.

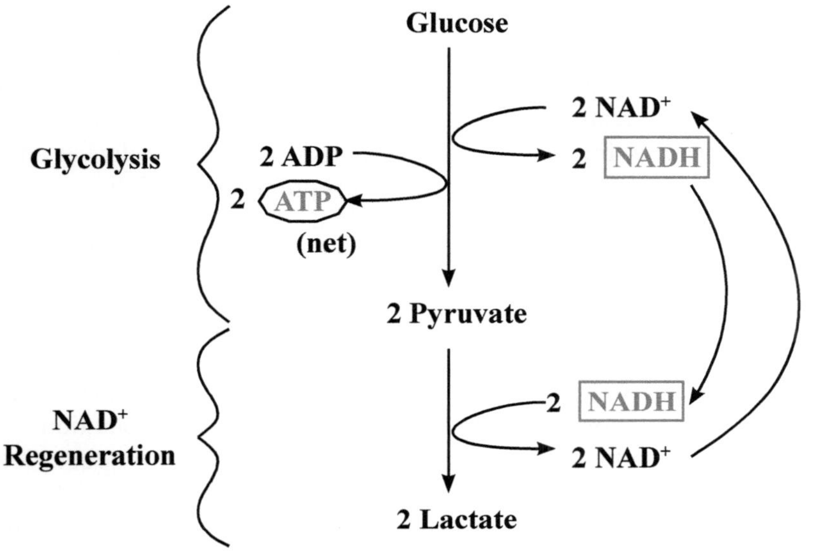

FIGURE 5.2 Genetic engineering, traditional mutagenesis, and strain selection.

Biotechnology, in contrast to organic chemistry, typically uses water as the reaction media. Volume productivity, also known as fermentor yield, is the primary component defining the viability of a process to handle product separation from biomass and water-soluble residues as well as product recovery from an aqueous environment. Any microbial process can now, at least theoretically, be improved by genetic engineering to provide industrial yields. For industrial enzymes and medicinal proteins, this is the easiest process to follow. Amino acid fermentation processes and the production of various secondary metabolites also benefit greatly from this. However, genetic engineering is unable to overcome the physicochemical constraints of fermentation processes, such as product inhibition or cell membrane damage brought on by high alcohol or acetic acid concentrations as fermentation products. The other requirements for a wider application of fermentation and biotechnology are therefore process engineering and the creation of new or enhanced recovery methods.

5.4.1 YIELD-DEPENDENT MANUFACTURING COSTS

The main contributor to an efficient fermentation process's reduced variable cost is the yield per carbohydrate substrate. The fixed expenses of the process, like as labor costs and depreciation on plant infrastructure per kg or ton of product, are instead determined by the fermentor yield and the fermentation time hatch cycle. Any increase in recovery yield has an impact on both variable and fixed expenses. The extraction efficiency per 1 culture broth is frequently in the mg or even pg range during an early research stage employing wild-type strains produced in a lab.

However, the production cost of bulk fermentation products like citric acid and ethanol is dominated by the carbohydrate substrate. The manufacturer must therefore look for a cost-optimum for bulk products in the current competitive environment that takes into account yield and process parameters on the one hand and actual carbohydrate prices, quality, sourcing reliability, and waste disposal logistics on the other. The most important economic metric for bulk product fermentations, from an industrial standpoint, is the weight per weight (w/w) product yield on substrate under large-scale production circumstances. Citric and LAs are good examples in this regard because they may be produced with high w/w yields of 85% and 98%, respectively.

In comparison to C_6-carbohydrate, other fermentation products including ethanol, butanol, and acetic acid suffer from a two mole CO_2 loss per mole loss. Their w/w yields are less than 50% on an industrial scale, which increases the economic impact of the carbohydrate substrate. Even if it needs some administrative support, bioethanol can successfully compete with its synthetic counterpart. A w/w yield of 50% from low-yield fermentation techniques cannot possibly produce acetic acid at the current cheap costs of an ethanol or methanol chemistry. However, the cost-competitiveness of fermentative acetic acid would soon rise if an efficient anaerobic homo-acetic acid fermentation could be developed, which would result in the synthesis of 3 moles of acetic acid per mole of C_6-carbohydrate while avoiding CO_2 as the principal by-product.

5.4.2 SUBSTRATE CHOICE

The genetic make-up of the microbe determines the chemical properties of the carbohydrates substrate, such as whether and which mono-, di-, or polysaccharide can be employed for a given fermentation process. Most organisms require glucose for growth, while many varieties can also hydrolyze sucrose. A P-galactosidase must be induced, and there must be a mechanism for moving lactose across the cell membrane. Similar to this, because starch lacks amylases necessary for starches liquefaction and saccharification, many key fermentation microbes cannot thrive on starch, the least expensive specified carbohydrate substrate available. For strains possessing enough of their own P-amylase or glucoamylase activity to convert dextrin into maltose or glucose, respectively, enzyme-liquefied starch is the preferred substrate. As previously stated, there are very few instances where the straightforward creation of the metabolite depends on a specific carbohydrate substrate, such as the production of gluconic acid from glucose. In all other circumstances, the genetic makeup of the producing strain rather than the chemical process leading to the desired end product dictates the choice of substrate. Simply put, fructose serves as the central entry door via which all carbohydrates are ultimately routed into the central metabolism. An empirical finding is that the best sugar or carbohydrate for growth is not always the best substrate for the creation of the best product. Higher solid yields and increased volume productivity are frequently provided by carbohydrates that restrict growth.

Similar to how different catabolic shunts affect how carbs are digested, they can have a significant impact on the overall product yield. The choice of substrates that are available for the specific production process may be further reduced as a result, and often, rather expensive carbohydrates may be preferred. The final substrate choice for commercial production is determined by an interactive ranking of quality, price, and technical and process aspects within the range of carbohydrates that the microbial strain will accept based on its genetic makeup. Refined glucose, sucrose, or lactose of food- or even pharma-grade is used in pricey cell culture media for the manufacture of monoclonal antibodies or bacterial fermentations for recombinant human proteins that require substantial downstream processing and purification. Irrespective of how important the price of raw materials is related to other costs, the cost of the carbohydrates finally rises to the level of serious concern during the growth phase and is closely watched during production for all other fermentation systems. The most overpriced sugar, sucrose, is a rare fermentation substrate as opposed to the less expensive residues of sugar refining, cane, and beet sugar molasses. Brazil uses beet sugar molasses for fermentation process, cane sugar molasses for ethanol production, and both molasses is still utilized for citric acid production.

Citric acid is a good substrate substitute having an excellent general dynamics for even complex fermentation processes. Low quantities of trace metal ions in the fermentation medium are necessary for high citric acid yields because they encourage citric acid buildup rather than further oxidation to CO2. By pre-treating the common raw material molasses with ferrocyanide, this is accomplished. Surface and submerged fermentation both require this pre-treatment; however, the former has mostly been abandoned due to overall equipment and manpower cost limits. Conventional citric acid recovery involves treating precipitated calcium citrate with dilute sulfuric acid, which produces free citric acid and results in the creation of gypsum. Gypsum deposition in landfills is becoming more and more challenging; its discharge to gypsum processors is significantly made more difficult by the material's dark blue hue as a result of the ferrocyanide treatment. By switching from molasses to glucose, a new recovery technique based on solvent extraction or ion-exchange chromatography can completely replace the generation of gypsum or provide a white gypsum byproduct that is more marketable. Raw materials, byproducts, and garbage that have not been refined. The economic advantages of less-refined agricultural raw materials and food processing by-products, such as whey or lactose crystallization leftovers, are frequently threatened by the inhibitory effects of the organic and inorganic contaminants or by exacerbated product extraction and purification. This is comparable to the molasses problem. By-products of agriculture and food processing, particularly in bulk fermentations, transmit waste disposal issues, including solid waste, BOD, COD, and salt freight, from the primary processor to the fermentation industry. Derivatives and waste materials must therefore be carefully considered in terms of overall cost savings while taking into account the additional processing and logistics challenges. There are numerous instances, though, where the complex impurities of corn steep liquor, molasses, or whey fractions can supplement the production strain's needs for amino acids, nucleotides, vitamins, or salt that would otherwise need to be met by the pure compounds or by still-expensive protein hydrolysates and yeast extracts.

5.4.3 REDUCING SUGARS AND MEDIUM STERILIZATION

By-product impurities frequently have inhibitory effects because they produce undefined Maillard products when they reduce sugars with nitrogen molecules. In addition to the potential reduction in yield, the Maillard products' brownish color, which necessitates more work during the purification process, must be taken into account. In medium sterilization, the production of Maillard products is another issue. The fermentation process frequently gets worse when glucose or lactose are sterilized jointly with nitrogen-containing medium components. In theory, this can be prevented by filter sterilization or by separately sterilizing the reducing sugar, neither of which is very practical for large-scale fermentation processes. Sucrose would be preferred due to its better sterilization stability, but only at competitive prices. Native starch cannot be used as a substitute since it gelatinizes when heated. Starch as a source of carbohydrates in agitated aerobic fermentations, the considerable viscosity of starch-containing media prevents the essential oxygen entry. The optimal carbohydrate substrate for all microbes that can use dextrines is liquified starch treated with conventional or temperature a-amylase before the sterilizing procedure. A further benefit is offered by the use of dextrines or liquified starch. It can be viewed as a reservoir or storage carbohydrate that slowly releases energy during fermentation. It has a low osmotic value, which reduces substrate inhibition and peak oxygen requirements that are typically seen in industrial production media with high mono- and di-saccharide concentrations.

5.4.4 INTEGRATED PRODUCTION COMPLEXES

Access to a feedstock supply will result in a cost advantage at a specific production volume per manufacturer and plant location. Regarding the manufacture of citric acid and the ambitious LA development projects, the starch and fermentation sectors are already exhibiting a strong forward and backward integration trend. In order to manage feedstock, byproduct, and waste disposal logistics, backward integration is much more crucial at the production volume and plant scale of the fuel ethanol makers. Based on the exact market situation for methanol, ethanol, and refined sugar, including the sugar export terms, the cane sugar sector in Brazil can direct varying proportions of simple sugar syrup, refining residues, and molasses into ethanol production. Bagasse is burned to produce energy when it comes to the waste products. Vinasse, a byproduct of distillation, is recycled through irrigation, a process that has been used for more than a century without causing environmental harm. All of the major fuel ethanol producers in the US are subsumed inside the corn wet-milling sector, or to be more exact, are a product of the agricultural sector. This corporate structure is unquestionably necessary for procuring feedstock that is cost-competitive as part of integrated management of all maize refining products. The distribution balance of starch, gluten, corn oil, and steep water to its developed markets may no longer be sustainable at a particular scale of operation, supplied that an additional million-ton bulk product may appear. This will promote the switch to the corresponding unrefined feedstocks, cane sugar juice, whole or shredded maize, and grain, in addition to the lower cost per kg of fermentable carbohydrates. Naturally, this has a direct impact on the mass streams of waste materials, including liquid and solid, as well as byproducts and feedstock. Since the bulk fermentation plant can be incorporated into the agro-industrial complex as part of a closed-cycle operation in an ideal setup, many challenges are then relocated from the feedstock level to the processing sector, as was previously indicated. Only bioethanol is now produced on a scale that makes the aforementioned factors relevant. Citric acid aside, all other current bulk fermentation products can at least still have their waste discharge issues resolved using well-established principles: liquid wastes to an aerobic or anaerobic sewage plant, solid wastes to a landfill or burning, and biomass to valorization or discharge as feed additives and fertilizers as alternative outlets.

5.4.5 NOVEL CARBOHYDRATES-DERIVED GOODS

The commercial market for numerous innovative fermentation products has grown during the past 10 years. The majority of items are high-priced pharmaceutical specialty, either recombinant mammalian proteins or natural microbial secondary metabolites. Unquestionably, the distinct medicinal uses of these products and the distinctive manufacturing method of transgene gene expression will support the continuing improvement and addition of new specialty fermentation – without the need for organizational interference, without a significant impact on resource management, and without any major impact on the value addition of agricultural surpluses. The success of the few new items that have been introduced into the technical markets has varied. Both poly-P-hydroxybutyric acid, a biodegradable polymer, and xanthan, a microbial thicken that competes with plant gums, suffer in large-scale applications due to their comparably high prices and production costs. For both products, processing challenges linked to the bacterial thickener's viscosity and the intracellular manufacturing of the bacterial polymer, respectively, place limitations on cost rather than fermentor yield. Fermentative surfactants are likewise expected to have similar processing restrictions. The limitations of fermentation technology in the production of functional polymers are emphasized, and they highlight newer possibilities such as enzymatic production from monomers or synthesis in transgenic plants. The two primary product approaches for organic compounds from fermentation—performance versus bulk—are well illustrated by LA. In the technical industry, d-lactic acid has found a niche use as a cheap chiral intermediate for enantiopure herbicides. The numerous optically active fermentation products have a great deal of promise to be used as chiral intermediates in the synthesis of contemporary low-dose medicines and insecticides. Many US starch makers prefer L-lactic acid as a novel bulk chemical derived from agricultural raw materials. Although the market for LA in foods is largely saturated, LA ethers and polymers are predicted to open up new technical markets for the conventional molecule. Competitive price for LA and compounds made from LA is a requirement for market access. Because substrate and fermentor yields of LA in batch or continuous culture are relatively high, this once more relates to downstream processing.

The old-fashioned calcium lactate/gypsum method could be replaced by direct esterification, distillation, and LA separation by electrodialysis, which are all affordable and theoretically feasible recovery technologies. The majority of prospects for new chemical products made from renewable resources are generally found in anaerobic fermentation. Propanol, butanol, propionic acid, butyric acid, propanediol, and butanediol are examples of traditional alcohols and acids that can operate as C_2–C_4-intermediates in well-established chemical reactions. These anaerobic fermentation products may see a resurgence as the cost and availability of common petrochemical feedstocks change. However, this is presuming that novel processing ideas as well as cutting-edge methods like genetic engineering and metabolic design can considerably improve their manufacturing processes. Future evaluations of chemicals produced from nature will surely also consider their eco-balance and life-cycle assessments. Unquestionably, energy net-gain and all other factors of direct and indirect pollution will be taken into account and used as competitive reasons. However, any technological and cost structure advancement will concurrently result in a better ecobalance of these traditional or novel fermentation items, according to the development theories and product policies of today.

KEYWORDS

- **biofermentor**
- **fermentations**
- **integrated production complexes**
- **microbial fermentations**
- **novel carbohydrates**
- **raw material**
- **sterilization**
- **yield-dependent manufacturing costs**

REFERENCES

Anaerobic Respiration, (2015). Wikipedia. https://en.wikipedia.org/wiki/Anaerobic_respiration (accessed on 18 November 2023).

Clostridium, (2016). Wikipedia. https://en.wikipedia.org/wiki/Clostridium (accessed on 18 November 2023).

Danovaro, R., Dell'Anno, A., Pusceddu, A., Gambi, C., Heiner, I., & Kristensen, R. M., (2010). The first metazoa living in permanently anoxic conditions. *BMC Biology, 8*, 30. http://dx.doi.org/ 10.1186/1741-7007-8-30.

Kimball, J. W. "*Blood.*" Kimball's Biology. https://www.biology-pages.info/B/Blood.html (accessed on 18 November 2023).

Kimball, J. W. "*Glycolysis.*" Kimball's Biology Pages. https://www.biology-pages.info/G/Glycolysis.html (accessed on 18 November 2023).

Lactic Acid Fermentation, (2015). Wikipedia. https://en.wikipedia.org/wiki/Lactic_acid_fermentation (accessed on 18 November 2023).

Lactic Acid, (2015). Wikipedia. https://en.wikipedia.org/wiki/Lactic_acid (accessed on 18 November 2023).

Loricifera, (2016). Wikipeida. Last modified. https://en.wikipedia.org/wiki/Loricifera#In_anoxic_environment (accessed on 18 November 2023).

Methanogenesis, (2015). Wikipedia. https://en.wikipedia.org/wiki/Methanogenesis (accessed on 18 November 2023).

OpenStax Biology. *Metabolism without Oxygen.* CC BY 3.0. Download the original article for free at http://cnx.org/contents/185cbf87-c72e-48f5-b51e-f14f21b5eabd@9.85 (accessed on 18 November 2023).

Raven, J. B., Urry, L. A., Cain, M. L., Wasserman, S. A., Minorsky, P. V., & Jackson, R. B., (2011). Cellular respiration and fermentation. In: *Campbell Biology* (10th edn., pp. 162–184). San Francisco, CA: Pearson.

Raven, P. H., Johnson, G. B., Mason, K. A., Losos, J. B., & Singer, S. R., (2014). How cells harvest energy. In: *Biology* (10th edn., pp. 122–146). AP. New York, NY: McGraw-Hill.

Robert, B., David, R., Bruce, S., Smith-Caldas, M., Christopher, H., & Eva, H. (2017). Metabolism without oxygen. In: *Principles of Biology*, by CC BY 4.0. Download the original article for free at http://cnx.org/contents/db89c8f8-a27c-4685-ad2a-19d11a2a7e2e@24.18 (accessed on 18 November 2023).

Sulfate-Reducing Bacteria, (2015). Wikipedia. https://en.wikipedia.org/wiki/Sulfate-reducing_bacteria (accessed on 18 November 2023).

Terry, R., (2012). *Anaerobic Respiration.* CK-12. http://www.ck12.org/user:dGVycnlyQHZhbGxleTI2Mi5vcmc./section/Anaerobic-Respiration-%253A%253Aof%253A%253A-Photosynthesis-and-Cellular-Respiration/ (accessed on 18 November 2023).

CHAPTER 6

Microbial Production of Metabolites

6.1 INTRODUCTION

Various primary metabolites from microbial origin are being commercially produced and widely used in food and feed manufacturing. These consist of amino acids, alcohols, organic acids, vitamins, flavor modulator nucleotides, polyols, polysaccharides, and sugars. In the previous three decades, conventional industrial microbiology has used modern techniques to improve the efficient production of these food and non-food products. Genetically improved microorganisms through recombinant DNA technology have gained specific interest in this field. The major microbial hosts for recombinant proteins are *Aspergillus niger*, *B. subtilis*, *E. coli*, *S. cerevisiae*, *Hansenula polymorpha*, *Pichia pastoris*, etc.

6.2 AMINO ACIDS

Numerous amino acids are commercially synthesized in large quantities through fermentation technology. This production is obtained through microbial biotransformation by high-yielding microbial strains. These are typically used as supplements in food and animal feed and flavor modulator additive. Although, various amino acids have also been extensively used in other industries such as pharmaceutical, cosmetic, and polymer chemical industries. Most important food application is of L-glutamic acid. Sodium salt of glutamic acid is predominantly used as flavor enhancer throughout the world.

6.2.1 L-GLUTAMATE

Among all amino acid commercially produced, L-glutamic acid is perhaps the most significant from quantity point of view. Monosodium glutamate also known as MSG, amplify and intensify the food without imparting its

Handbook of Industrial Food Microbiology. Manju Nehra & Vikash Nain
© 2025 Apple Academic Press, Inc. Co-published with CRC Press (Taylor & Francis)

own taste and widely used as flavor enhancer. It is naturally present in some foods and has been found to be the active component responsible for flavor-enhancing in seaweed stock that widely used in Far Eastern foods. In 1908, MSG was isolated from the seaweed *Laminaria japonica.* Japan started its commercial production in Japan immediately, later they used soya and wheat protein extracts for the production of this compound. As of its safe use history, in 1959, the US Food and Drug Administration (FDA) categorized it as generally regarded as safe (GRAS). In 1970 the Joint Food and Agriculture Organization (FAO)/World Health Organization (WHO) Expert Committee on Food Additives issued the guidelines for its acceptable daily intake as 0–120 mg/kg body weight.

MSG was developed commercially by costly and proteins hydrolysis based chemical processes for first 50 years. Acid hydrolyzed proteins generate a mixture of all the amino acid and isolation of glutamate from these hydrolytes make it more and more costly. Later, chemical synthesis contributed significant amounts of commercially produced MSG. But this process was also expensive because it produced a mixture of both D and L-glutamate. From this mixture, D-isomer had to be removed because it is tasteless.

In 1957 Kinoshita, Udaka, and Shimono of the Kyowa Hakko Kogyo Company discovered the glutamic acid fermentation in Japan. It is exciting that the founder of the company (Dr, Benzaburo Kato), concerned about Japanese community's diet and working to limit their overreliance on rice, had instructed Kinoshita with the task of developing edible proteins through fermentation process. The research teams instead came up with a fermentation process that generated an extracellular L-amino acid from non-protein substrate. This scientific breakthrough has been incredibly importance for the amino acid fermentation industry to develop amino acid commercially at lower cost.

MSG was produced using distinct species of the *Corynebacterium* and *Brevibacterium* genera. *Corynebacterium glutamicum* based overproduction (or fermentation) of the MSG was reported initially. Later, Numerous organizations comes up with various glutamate-producing microorganisms, for example, *Brevibacterium* and *Arthrobacter,* but all these strains were found closely related to *C. glutamicum.* These include various species of the closely related genera of *Corynebacterium, Brevibacterium, Arthrobacter, Microbacterium*, and *Micrococcus*. Generally, these are non-motile, biotin requiring Gram-positive bacteria having significant glutamate dehydrogenase activity. For most of industrial fermentations, the species

Brevibacterium and *Corynebacterium* are used. For example, the wild-type *Corynebacterium glutamicum* expresses feedback inhibition effect when dry weight concentrations of cellular glutamic acid reach up to 5%. Nevertheless, both the regulatory and auxotrophic mutant strains are the developed through mutagenesis and selection programs. Such strains were produced with a large concentration of cytoplasmic amino acid at a fast steady-state rate of production. They generate around 30% L-glutamic acid with a higher yield of 1 mol glutamate per 1.4 mol of glucose. Further, they are also found to be phase resistant. Recently, recombinant DNA technology has been used to improve the capability of particular biosynthetic enzymes by replacing these enzymes with multicopy plasmids holding the structural genes responsible for such production.

The general plan for attaining amino acid overproduction involves:

- Stepping up anabolic enzyme activity.
- Manipulating the regulation to eliminate feedback-based control mechanisms.
- Blocking pathways which lead to the production of undesirable by-products.
- Blocking pathways which degrade the target product.
- Controlling the ability to process the immediate precursor of l-glutamic acid, i.e., application of mutant strains that lacks oxoglutaric acid dehydrogenase. These mutant strains develop key intermediate substances from isocitrate through the glyoxylate cycle in the growth phase.

Therefore, as these bacteria usually not secrete glutamate, a variety of medications are used to render the cells more permeable and to further release the amino acid in the medium. These treatments include limitation of biotin, limitation of phospholipid biosynthesis by supplementing saturated fatty acids C_{16}–C_{18} during the growth phase, and integration of surfactants (for example, Tween 40) and antibiotics (penicillin) in the production media.

6.2.1.1 *INDUSTRIAL PRODUCTION OF L-GLUTAMIC ACID*

Industrial fermenters are usually stirred-tank reactors of stainless steel with up to 450 m^3 capacity. These are batch processes which are aerobically operated at 30–37°C, the specified temperature may vary depending on the microorganisms used. Media composition should include 121 g of glucose, 5 g of ammonium acetate, 6 g of molasses, 1.2 g of potassium hydrogen phosphate, 6 μg of potassium sulfate, 6 μg of manganese sulfate and 0.1 ml of antifoaming agent per liter of distilled water. Apart from sources of carbon and nitrogen, the fermentation medium usually includes inorganic salts, including magnesium, manganese, phosphate, and potassium, and reducing biotin amounts.

Nutritionally fastidious corynebacteria may also require other amino acids, vitamins, purines, and pyrimidines. Carbohydrates, ideally glucose or sucrose, are the preferred sources of carbon. Cane or beet molasses can be used, but the medium needs more alteration because they appear to be quite rich in biotin content. This may be solved by incorporating saturated fatty acids, penicillin, or surfactants that facilitate excretion. The source of nitrogen (urea, ammonia or ammonium salts) is gradually fed to avoid the suppression of L-glutamate development. By adding alkali, medium pH is maintained at 7–8, otherwise the pH gradually decreases with the production of L-glutamate into the medium.

L-glutamic acid accumulation in medium does not become evident until half-way through the fermentation which usually persists for 35–40 hrs. A concentration of 80 g/L L-glutamate is achieved through this process. Recovery of a product (L-glutamate) requires the separation of the cells from the culture medium. Crystallization of L-glutamic acid from the consumed media is carried out by using hydrochloric acid to lower the pH to its isoelectric point of pH 3.2. The l-glutamic acid crystals are then filtered off and washed away. MSG is prepared by applying a sodium hydroxide solution to the l-glutamic crystalline acid accompanied by recrystallization.

6.2.2 L-LYSINE

Humans and other animals are not able to synthesize L-lysine. They have to obtain this essential amino acid as part of their diet. Most of cereals and vegetables in are also deficient in lysine. Food derived from these sources must be supplemented with L-lysine amino acid. To meet these requirements, around 4 lakh tons of L-lysine is commercially produced annually throughout the world. One fourth of this amount is currently produced through microbial fermentation and biotransformation process. Chemical synthesis contributes the remaining portion. Like L-Glutamate, this method produces a mixture of D- and L-isomers but only L-lysine is utilized by our body. Therefore, after chemical synthesis, optical resolution is needed while microbial development has the benefit that only the L-isomer is produced. Usually, *Entribacter aerogenes* and *E. coli* are employed for the microbial production of L-lysine.

6.2.2.1 INDUSTRIAL PRODUCTION OF L-LYSINE

Figure 6.1 represents the metabolic pathway of L-lysine development in wild-type *C. glutamicum*. Double ended products of branched pathway of lysine and threonine regulate the first major step of this metabolic pathway, aspartate to aspartyl phosphate, catalyzed by aspartokinase, through feedback inhibition. Homoserine dehydrogenase activity is also subject to feedback inhibition by threonine and repression by methionine. However, lysine accumulation, which is unusual for the first enzymes following the branch point of a pathway, does not inhibit dihydropicolinate synthetase. The over-producing strains of *C. glutamicum* selected for lysine production have defects in these feedback control mechanisms. They lack the activity of homoserine dehydrogenase and are therefore auxotrophic homoserins. These auxotrophs transform all aspartate semi aldehyde into lysine, as no feedback control due to the absence of threonine formation. However, to maintain the growth of auxotrophic bacterium, precisely calibrated quantities of isoleucine, threonine, and methionine have to be introduced into the culture medium. The majority of commercial L-lysine fermentations in aerated stirred tank reactors are operated as batch processes. Cane molasses is the most recommended source of carbon, although other carbohydrates, ethanol, or acetic acid may be used, usually accompanied by hydrolysates of soya bean. pH of the medium is maintained at neutral by adding urea or ammonia. These components also act as source of nitrogen in medium. The temperature of inoculum is maintained at 28°C for optimum fermentation. Biotin level control is critical since concentrations below 30 mg/L consequence in production of L-glutamate rather than L-lysine. However, cane molasses generally has enough biotin to meet this demand. By using a high inoculum concentration, generally about 10% (v/v) of the fermentation volume, the lag phase can be significantly shortened. Lysine production begins in the early exponential phase and continues in stationary phase. Approximately 60 h fermentations process yield 40–45 g/L l-lysine from a medium having 200 g/L concentration of molasses with 100 g/L sucrose. Recovery of lysine is relatively easy and simple. Upon separation of cells from the fermentation medium, it is acidified with hydrochloric acid to pH 2.0 and the L-lysine is entrapped in the form of ammonium on a cation-exchange column. Then, a diluted ammonia solution is used to elute out the product from the column. This eluate is then reacidified and the product is eventually crystallized as hydrochloride L-lysine.

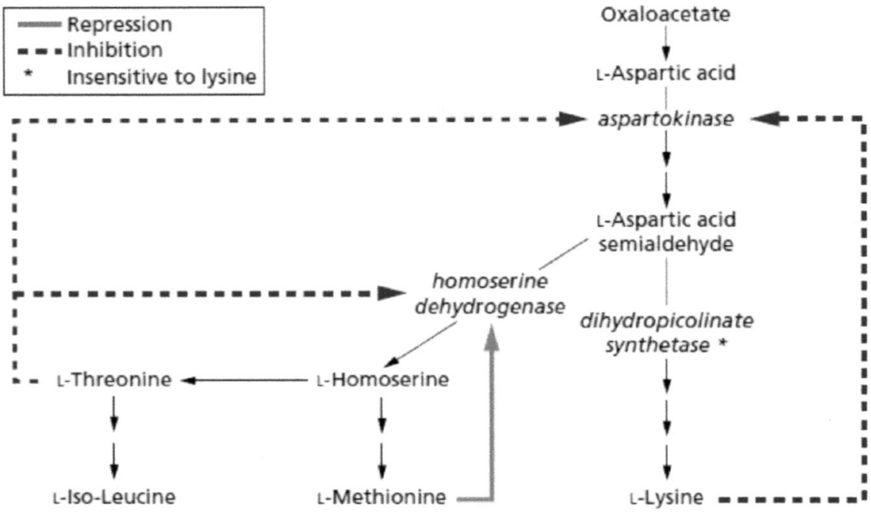

FIGURE 6.1 Metabolic pathway of biosynthesis of L-lysine.

Source: Reprinted with permission from Waites et al. (2009). Copyright 2009. John Wiley & Sons.

6.2.3 CITRIC ACID

Most important chemicals produced through the process of fermentation are organic acids which have a prominent role in biotechnological synthesis of various chemicals. Among them, citric acid has attracted significant attention due to its diverse utilities in food, cosmetic, pharmaceutical, and chemical industries. Its first commercial production was done by John and Edmund Sturage company in United Kingdom in the year 1826. In 1893, Wehmner observed the microbial occurrence of citric acid in the *Penicillium* and *Citomyces* culture. Millard in 1922 reported the production of citric acid by *Aspergillus niger* under the nutrient deficiency conditions. Pfizer starts the microbial production of citric acid in 1923 at commercial level. Significant quantities of citric acid are produced in large-scale industrial processing. With continuous increasing demands, industries are looking for inexpensive, profitable, and innovative process technologies for the production of organic acids.

6.2.3.1 MICROBES USED FOR CITRIC ACID PRODUCTION

Numerous microorganisms such as *Aspergillus* spp. (e.g., *A. niger* and *A. foetidus*), yeasts (e.g., *Yarrowia lipolytica*), and a few bacteria have been found to have citric acid-producing ability. James Currie identified and isolated a filamentous fungus as *A. niger* which can act as a citric acid-producing cell factory. But most of them do not meet the industrial sustainable production rate. This could be explained by the fact that citric acid is a metabolite of energy metabolism and only under drastic imbalances conditions there is an increase in its accumulation in considerable concentrations.

Niger is a very interesting and prominent host for citric acid production having many advantages as compared to other microorganisms such as ability to use cheaper raw materials as substrates, high production yield and easy handling. Additionally, the production of citric acid using yeasts has certain advantages, such easy genetic management, resistance to high substrate concentrations, ability to tolerate high metal ion concentration. Hence, the production of citric acid using *Y. lipolytica* has attracted considerable interest. To further reduce the manufacturing cost of bio-production of citric acid using *A. Niger*, *Y. lipolytica*, various biotechnological tools have been employed to increase the yield of citric acid such as effective mutagenesis, metabolic engineering and optimal fermentation strategies.

Altered fermentation methods by the same strain can result in different yields of citric acid production. Thus, a high yielding strain in solid state or liquid surface fermentation may not be equally good for submerged fermentation. So, each commercially important method and effect of raw material should be well studied for every known citric acid producing strains. In all citric acid production methods, microorganism inoculation is performed by means of spores introduced into the fermentation broth. Inoculum of spore can be done by either mixing the spore suspension or by incorporation of air administrated in medium. The spores are produced at optimum temperature in glass bottles on solid substrates. Citric acid production and spore viability by *A. niger* mycelia are affected by type of medium used for sporulation and incubation time.

6.2.3.2 SUBSTRATE

Numerous raw materials have been used for the industrial production of citric acid by submerged fermentation such as hydrocarbons, starchy materials and molasses, however it is commonly produced by submerged fermentation using starch or saccharose rich media. Citric acid is commonly produced by the fermentation using relatively cheap raw material, including natural raw products such as starch hydrolysate, broth of sugar cane, and by-products such as cane sugar and molasses of beet. Because of its relatively low cost and high sugar content (40–55%) molasses is usually used as the sugar source for industrial citric acid production, molasses quality varies significantly as it is a by-product of sugar industries. Due to this variation in quality, all types of molasses are not suitable for the production of citric acid. Various factors affect the molasses composition such as type of cane and beet, crop cultivation methods, pesticides and fertilizer used, handling and storage conditions, procedure of production, etc. Both beet and cane molasses are appropriate for the production of citric acid. Other agricultural residues also have been explored as substrate in solid stale fermentation for the production of citric acid. These include sugar beet cosset, kiwi fruit peel, cassava bagasse, pineapple waste, apple pomace, coffee husk, wheat, and rice straw, etc. It has been a growing trend towards efficient use and adding value to these residues, in addition to being a way to reduce environmental concerns.

6.2.3.3 CITRIC ACID PRODUCTION TECHNIQUES

The most economical and widely used way of obtaining this product is the synthesis of citric acid by fermentation. Around 90% of the global production of citric acid is obtained through fermentation, which has various advantages over other method. These advantages are simple and stable operational process, less requirements technical skills and sophisticated control systems, simple plant organization, low energy consumption, etc. The production of citric acid by fermentation can be divided into three stages which include raw material preparation and inoculation, fermentation, and product recovery. The production of industrial citric acid can be carried out in three different ways: submerged fermentation, surface fermentation, and solid-state fermentation (SSF) (Koji process). Raw material and inoculum preparation is required in all three methods.

1. **Submerged Fermentation:** This technique is widely applied to the production of citric acid. It is assessed that this technique accounts for around 80% of world production of citric acid. This industrial-level fermentation process requires more sophisticated equipment and laborious control. On the other side, it offers several benefits such as

high yields and productivity, lower labor cost, lower contamination risk and need of labor. Submerged fermentation may be performed in batch, continuous or feed batch processes however the batch method is used most commonly. Based on processing conditions, citric fermentation is usually completed within 5 to 12 days.

2. **Surface Fermentation:** Citric acid production process is first carried out by liquid surface production method. This technique was used for commercial production until the process of submerged fermentation was adopted. Surface fermentation is still used in medium- and small-scale industries because it involves lower energy, installation and operating costs.

   The process is conducted in fermentation chambers, where various trays are stored in racks. In these shallow trays, culture medium is stored with a capacity of about 0.4 to 1.2 m^3. Fungus grows in the form of mycelium mat on the surface of the medium. These trays are made up of polyethylene, highly pure aluminum wi,th special grade steel. However, citric acid yield was found higher with trays made up of steel. Proper aeration facilities are provided in fermentation system to maintain the temperature and humidity. Evaporative cooling is energy efficient and performed by passing air over the surface of fermentation chamber. Air supplied for this purpose should be sterilized so passed through bacteriological filters especially for the initial two days of spore's germination. *Penicillia* and *Aspergilli*, with some LA producing bacterial mos,t commonly contaminate the medium. During fermentation period (8–12 days) appropriate aeration is required to control the temperature as heat is produced upon fermentation. Then, tray contents are separated into crude fermentation fluid and mycelial mats. Mats are washed to remove the flooded citric acid.

3. **Solid-State Fermentation (SSF):** Or Koji process was first developed in Japan. Japan has plentiful availability of raw materials for such fermentation such as rice bran and fruit wastes. This is the easiest approach for the processing of citric acid and became an alternate process for the utilization of agricultural residues. Solid-state culture is defined by the production of microorganisms in an environment of low water activity on an insoluble material which acts as both nutrient source and physical support. Due to some growth of fungus on surface of the raw materials this process has some resemblance

with surface fermentation process. Moisture content of solid substrate is raised up to 70% based on the absorption capacity of substrate. The material's initial pH is usually set to 4.5–6.0, and the incubation temperature is maintained about 28–30°C, according to the microorganism used for fermentation. Fermentation was completed in about 96 hours when incubated at optimum conditions. *Aspergillus niger* is the most commonly used microorganism for solid-state fermentation. Although, few studies were also carried out with yeasts. The strains with large requirements of nitrogen and phosphorus are not ideal microorganisms for solid culture due to lower diffusion rate of nutrients and metabolites occurring at lower water activity in solid-state process. Pre-treatment of the substrates is not obligatory in this method. This is one of the significant benefits of solid culture.

6.2.3.4 PRODUCT RECOVERY

Citric acid recovery from fermented broth is generally done by three procedures: precipitation, solvent extraction, and adsorption. Precipitation method is the most frequently used, and applicable for all processes. The second method (solvent extraction) is applicable for fermented broth with very small impurities. Filtration based separation of fungal micelles and suspending material from fermented broth is required in these both methods.

The process of precipitation is the conventional procedure and achieved by introducing calcium oxide hydrate. Citric acid is converted into less soluble tri-calcium citrate tetrahydrate. These precipitates are collected by the process of filtration after treating it with sulfuric acid which form calcium sulfates. Mother liquor having citric acid solution is treated with activated carbon and transferred through ion exchangers of both cation and anion. Filtered liquor is then concentrated at 20–25°C temperature in a vacuum crystallizer to form citric acid monohydrates. If temperature is kept higher than 36°C, then it will yield the anhydrous citric acid. These crystals are collected by centrifugation and dried in bed dried for storage.

Solvent extraction is another approach used to purify and crystalize the citric acid. Generally, a combination of n-octyl alcohol, tridodecylamine, and isoalkane is used in this process. Many other solvents such as methanol, acetone, and ethanol were also explored for the extraction of citric acid from SSF processes. Isolation of citric acid by liquid-liquid extraction method has proved to be a realistic alternative to traditional techniques. Various amine- and phosphorous-based oxygen-having extractants are used for this purpose. The principal issues in the processing of citric acid are still the steps involved in the isolation and purification. Cost of citric acid production become high due to multistage filtrations, crystallization, precipitation, and drying process.

Several techniques have been recognized, in recent years, to reduce the recovery expenses, aiming to resolve the limitation of the precipitation process, which is associated with environmental pollution through the production of large quantities of calcium sulfates. Electro-dialysis is a method of electro-chemical isolation in which electrically charged membranes and the difference in electrical potential are used to isolate ionic substances from aqueous solutions. This method was evaluated in the recovery of citric acid and demonstrates more economical isolation of ionic compounds from filtered fermentation broths. The concern is that the expense of the procedure of electrodialysis was found to be around 50% higher than the conventional method of recovery of citric acid on an industrial scale. The higher costs of electromembrane and electrical power appear to restrict its implementations to high value-added products only.

6.2.3.5 APPLICATION

Citric acid is extensively used in food industries to impart a pleasant tart flavor to various food commodities such as candies, beverages, dairy products, jam and jellies. There are a lot of non-food applications of citric acid. These include detergent additives, pharmaceuticals, cosmetics, etc. High water solubility, pleasant taste, ability of buffering and chelating, and safe for human consumption make it ideal for the pharmaceutical and food industries. It is also widely used in dishwashing cleaners and household detergents as a co-builder with zeolites, predominantly in concentrated liquid detergents.

6.3 ENZYMES

6.3.1 COBALAMIN (VITAMIN B12)

Vitamin B_{12} is often used as a dietary supplement and is particularly essential in the diagnosis of pernicious anemia. It is obtained industrially through two stage fermentation mainly assisted with bacteria *Propionibacterium shermanii* or *Pseudomonas denitrifican.* Like many other fermentation processes, its microbial production is also controlled by feedback mechanism. First stage of fermentation for cobalamin proceeds in anaerobic condition. 5,6-dimethyl benzimidazole is precursor of vitamin B_{12}. First-stage fermentation is performed essentially in the absence of this precursor. This prevents the vitamin synthesis in first stage otherwise it will produce repressive effect and inhibit the further production of cobalamin. Absence of precursor led to the accumulation of an intermediate compound known as cobinamide. In second and aerobic fermentation, culture is mixed with dimethyl benzimidazole and aerated to convert the intermediate compound to vitamin B_{12}. Cyanocobalamin is obtained in isolated form upon purification.

6.3.2 RIBOFLAVIN (VITAMIN B2)

Processed foods, especially breakfast cereals and soft drinks, are fortified with riboflavin. It's made either by chemical processing or by fermentation method. Its fermentation process commonly used yeast stains are *Eremothecium ashbyii.*, *Ashbya gossypii*, and *Candida flareri.* Production of riboflavin through fermentation is mainly carried out by utilizing the *A. gossypii* strain, and usually 600 tons of riboflavin is produced annually by this method. *A. gossypii* strains have higher stability as compared to other yeast used in riboflavin production. These strains have been designed to yield 10–15 g per liter riboflavin in aerobic fermentation at 34–37°C and pH 6.5–7.0. Here, fermentation process is basically divided in to two steps. First step encourages the growth of the microbial culture whereas second step enhance the production of riboflavin in fermentation process. This is accomplished by limiting the growth by governing the addition of a carbon source or a micronutrient such as iron on a consistent or continuous basis. Although, most of the riboflavin is secreted into the medium, yet some parts remain attached to the cells and that must be released by thermal treatment before further purification.

Riboflavin may also be produced from a *Bacillus subtilis* strain that has been genetically engineered to grow rapidly with high production rate. To deregulate riboflavin synthesis, conventional genetics technique was used, while genetic engineering strategies were employed to incorporate effective promoters and to enhance the copy numbers of the genes that are responsible for encoding of the enzymes associated with riboflavin biosynthetic pathways. The riboflavin obtained is similar to that obtained by conventional chemical methods, and it is allowed for use in food and feed in some countries. The UK Government, for example, approved the recommendation of the Advisory Committee on Novel Foods and Processes in 1997 that this commodity be granted food safety clearance.

6.4 BAKER'S YEAST

Without any doubt, yeasts are the most significant class of microorganisms currently being commercially used and consumed safely by humans. More than a century ago, bakers' yeast was first commercially produced in Europe and Northern America. *Saccharomyces cerevisiae*, also known as baker's yeast, is the most natural food-grade yeast, and it's used to make bread and other baked goods all over the world. *S. cerevisiae* is the most commonly utilized yeast genus, with specified strains used to make beer, wine, distillates, and ethanol in breweries, wineries, and distilleries. Molasses, a by-product of sugar industry, is used as a raw material in the production of baker's yeast. Commercially, *S. cerevisiae* and other yeast derivatives are abundantly available in the market to meet the requirements of the bakery and beverage sectors, as well as for food supplements and feed for human and/or animals.

Depending on the form and growth factors, fresh baker's yeast contains approximately 30–33% dry materials, 6.5–9.3% nitrogen, 40.6–58.0% proteins, 35.0–45.0% carbohydrates, 4.0–6.0% lipids, 5.0–7.5% minerals, and various amounts of vitamins. Commercially available fresh baker's yeast comes in liquid, creamy, or compressed versions, and also in the form of active dry yeast. The most widely used product is compressed baker's yeast, which is made up of single yeast species, i.e., *S. cerevisiae*. Specified strains of *S. cerevisiae* are used in the development of the instant dry yeast and active dry yeast. Instant dry yeast is normally in the form of fine particles which don't require rehydration before use, while active dry yeast is made up of grains or beads of live and dried yeast cells with a suitable leavening potential.

Contrary to active dry yeast, inactive dry yeast has no leavening properties and basically used in the dough conditioning to improve its baking properties and for the development of characteristic flavor.

6.4.1 NUTRITION

Baker's yeast has a strong metabolic activity in yeast cells as well as a large cell density. As a result, the yeast industry requires not only effective biomass growth (because yeast is a product component), but also greater cell mass with at a higher level of metabolic activity for the leavening operation. For the yeast industry, molasses is the single most important source of nutrition for yeast production. Under ideal growth conditions, 2 g of sucrose is needed to produce 1 g of bakers' yeast (dry basis). Since molasses includes 50% sucrose, 4 g of molasses produces 1 g of bakers' yeast (dry basis). Molasses is the preferred medium for the production of baker's yeast. As a carbon source, cane or beet molasses are used, depending on supply and economics. Alternative carbon sources include cheese whey and corn syrup. The nitrogen supply for commercial baker yeast cultivation is ammonium salts, aqua ammonia, or urea. Baker's yeast can also absorb L-forms of amino acids. Most yeast plants, however, tend to use basic aqua ammonia or ammonium salts due to production cost considerations. Six major B complex vitamins have been identified as necessary for yeast production. Biotin, pantothenic acid, inositol, thiamine, pyridoxine, and niacin are the six required vitamins. Biotin, an important vitamin for baker yeast, is deficient when the yeast is grown solely on beet molasses. Biotin levels in cane molasses are usually sufficient to sustain yeast production. Thiamine, which is enriched in growth media, is believed to speed up the fermentative processes of baker's yeast. The propagation of active yeast needs inorganic elements including potassium, magnesium, sodium, sulfur, iron, copper, and zinc. Yeast is usually given phosphoric acid as a source of phosphorus. In yeast cells, metaphosphate is believed to be a source of phosphate.

6.4.2 YEAST STRAIN

For industrial processing, baker yeast strains are stored as pure colonies, typically on agar slants in test tubes. Well-known pure culture techniques are commonly used to transfer them. Such plants are used commercially to initiate the fermentation process for this purpose. Baker yeast strains used in industry are very stable. Baker's yeast is top fermenting yeast that belongs to the *Saccharomyces cerevisiae* species. This class contains a large number of strains. All have very different characteristics, and yeast producers go to considerable lengths to identify the strains that closely fit to the various local requirements.

In industrial yeast development, proliferation is vegetative in nature, however *Saccharomyces cerevisiae* is a spore producing yeast, and new strains have been produced from various spores to develop yeast with better attributes (French Patent No. 13363631961). Yeast strains used these days can be classified into two categories. First one has the enzyme mechanism for maltose fermentation, making it ideal for sugar-free dough. The other one that have ability to ferment in a lag phase in dough made from wheat flour and yeast, lacking other fermentable sugars. This yeast, on the other hand, has a higher osmatic tolerance and better keeping quality, and is hence used in a wider part of world. *Torulaspora delbrueckii* (syn. *Saccharomyces rosei*) has been frequently used in Japan, and a hybrid variant of *Saccharomyces cerevisiae* and *T. delbrueckii* have been produced in Germany.

6.4.3 PRODUCTION METHOD

The single-cell culture system invented by Emil Chr. Hansen at the Carlsberg Laboratories over a century ago is the foundation for yeast cultivation. The starting medium is a laboratory-scale pure culture. The culture is inoculated on a sterilized material and allowed to mature until it produces a suitable volume of absolutely pure yeast. This is transferred from the laboratory to the propagating plant's first fermenter. A substrate, composed of a molasses solution to which different salts and anti-foam agents have been applied, is prepared in the laboratory first. Before adding the laboratory culture, the substrate is sterilized and cooled. After about 20 hours of light aeration, the mixture is moved to the next propagation point (ideally a fermenter with 15,000 litter capacity). Substrate must be sterilized and cooled in a fermenter before the end product of first stage is pumped in it. Light aeration is used once again, and the fermentation process is completed after approximately 16 hours. At this point, the contents are shifted to one of the large fermenters. The two propagating tanks described above are usually kept at different places, where all measures can be taken to prevent microbial contamination, and the substrates (water and air) can be sterilized before use. This produces a stage 2 fermented liquid that is usually free of contamination. The yeast is then further cultivated in large, closed fermenters with air injection and continuous substrate feeding for the entire duration of the cultivation.

6.4.4 APPLICATIONS

Bakers' yeast has two main functions: leavening baked goods (breads, Danish pastries, brioche, and croissants) and imparting fermentative fragrance and flavor in bakery products. Baker's yeast is primarily used as an important baking component, producing fermentation in the dough used to make bakery goods. This procedure facilitates in the production of soft and fluffy bakery products such as breads, bread rolls, pizza bases, cracker, cookies, sweet breads, and burger buns, etc. Yeast has been used in xylitol biotechnology because of its beneficial physicochemical characteristics. Baker's yeast has long been used as a biocatalyst in organic synthesis owing to its low cost and easy accessibility. Baker's yeast contains enzymes that can efficiently reduce a carbonyl group to a hydroxyl group, making it ideal for syntheses in organic compounds. Since it is readily available, cheap, and simple to handle, baker's yeast is widely used as a biocatalyst in various organic reactions such as oxidation, reduction, condensation, hydrolysis, cyclization, and so on. Yeasts, especially *S. cerevisiae*, are progressively used as hosts for the expression of protein biocatalysts and multi-enzyme pathways for the synthesis of fine chemicals and small molecular weight compounds of medical and nutritional significance.

6.5 ACETIC ACID

Acetic acid (CH_3COOH) is a colorless weak organic acid which have pungent smell and sour taste. Acetic acid is a primary raw material within the chemical industry. It is widely used for commercial production of many chemicals such as acetic anhydride, esters, and monomer of vinyl acetate. Approximately 75% of the chemical industry's acetic acid is produced by carbonylation of methanol by carbon monoxide. The remaining 10% is synthesized by microbial fermentation and mainly obtained as vinegar. Industrial acetic acid development was carried out exclusively by the transformation of sugar till the late 1800s when wood distillation process became competitive with fermentation method. After this, Petroleum acted as primary source for the production of synthetic acetic acid. In 1916, the first chemical unit was built for the manufacture of acetic acid. Acetic acid produced from corn fermentation was usually utilized in the food manufacturing. More than 8 million ton per annum acetic acid is produced worldwide through these processes (fermentation and synthetic).

6.5.1 *MICROORGANISMS INVOLVED IN THE FERMENTATION*

Acetic acid bacteria of the family Acetobacteraceae are categorized into 19 genera. Majority of these are obligate aerobes have an ability to grow on alcohol/sugar rich substrates. Acetic acid bacteria possess a unique metabolism which makes them better from all other bacteria in the production of acetic acid. These bacteria have an extremely high potential to oxide the various carbohydrates such as sugars and their alcohols and ethanol into acetic acid. The species engaged in development acetic acid are commonly grows on the top surface of the raw material and develop a jelly like layer. Industrially, this layer is also called "mother of vinegar." It is basically made up of a combination of yeast (*Saccharomyces cerevisiae*) and bacteria (*Acetobacter acetic*). Maintaining an acidic environment is necessary to suppress the growth of undesirable species and to promote the existence of bacteria generating valuable acetic acid. While there are a variety of bacteria which are capable to produce acetic acid, only a few are often used commercially (*Acetobacter*, *Gluconacetobacter*, and *Gluconobacter*). It is comparatively easy to isolate acetic acid with fungus as compared to bacterial production. Agricultural or fruit waste, hydrolyzed soya flour whey lactose, corn steep liquor and filtered silage of ethanol are the main substrates used for the cost-effective production of acetic acid.

6.5.2 ACETIC ACID PRODUCTION

The three popular approaches used in the production of acetic acid are trickling method, the submerged fermentation and the Orleans traditional method. Traditionally natural and spontaneous fermentation is also practiced in some area. The trickling or generator method is the quickest one and is commonly used in large scale vinegar production.

6.5.2.1 THE ORLEANS METHOD

One of the oldest and well-established approaches for the production of vinegar is the Orleans method. It is a steady and continuous operation, emerged in France. As a starter culture, high-grade vinegar is used to which wine is added at an interval of one week. This fermentation is carried out in high-capacity barrel (200 liters). Along with 15 liters of wine, about 65 to 70 liters of high-grade vinegar are introduced to the barrel. An additional 10–15 liters of wine are introduced after one week, and this is replicated at periodic intervals. Wooden barrels are loaded with liquid fermenting alcohol to around 3⁄4 of maximum volume and supplied with holes only a few inches above the liquid surface at the end of the barrel. These holes are kept opened, with a fine screen cover. Vinegar should be extracted from the barrel after around four weeks (10 to 15 liters a week), while additional wine is used to substitute the vinegar. Major problem associated with this method is periodic addition of wine without disturbance of floating bacterial mat. It can be done by using a glass tube that stretches up to the barrel bottom. Additional wine is pumped through the tube, not disturbing the bacterial mat. Sometimes, wood shavings are attached to the fermentation barrel to effectively protect the bacterial matter.

6.5.2.2 GENERATOR PROCESS

In 1832, German chemist Schutzenbach invented the trickle method (now known as generator fermentation or quick process) for producing vinegar in the early 19th century. In this process, bacteria were grown and produced a thick slime coating around a non-compacting material such as beech wood shavings, charcoal, or coke. Fermentation was more often accomplished in wooden or steel tanks with a capacity of 50,000–60,000 liters. A packaging material in the tank, on which the bacteria were immobilized, enhanced the surface area where bacteria were exposed to oxygen. The most often used packing material was made from beech wood shavings, on which the fluid was dispersed and then permitted to trickle down to the reactor's base, while the air was blown from the bottom part to ensure a high oxygen supply. The half-finished vinegar accumulated at the bottom of the tank was re-circulated to the surface until the required acetic acid concentration was achieved. This process is carried out at a temperature of 27–30°C, with a cooling coil in the tank to avoid overheating. The generator method had certain drawbacks, such as a high chance of clogging caused by cellulose-producing bacterial grown in the generator, the aggregation of dead bacteria, and contamination with vinegar eels. Another downside is the comparatively high loss of ethanol due to evaporation, which rendered producing vinegar with high acetic acid content impossible.

6.5.2.3 SUBMERGED FERMENTATION METHOD

Submerged culture was the most popular form of processing, since it improved general fermentation conditions such as aeration, stirring, and heating. Submerged culture fermenters were commonly used at industrial scales because generator culture was slow and costly. The mash is constantly stirred and aerated throughout this step, and the fermenters are equipped with a heat exchanger to ensure that the optimal temperature is maintained in the fermentation operation. Stainless steel fermenters with a variety of sizes are used basically, but mostly in the 10,000–40,000-liter range. Basic principle is that bacteria are free of substrate and air circulation in the medium by a stirrer placed at base of tank. The fermenter has mechanical agitation, which generate fine bubbling in the device. An air outflow, thermometer, cooling coils, and a device to monitor and eliminate foams is located at the top of a fermenter. This system is extremely sensitive, and since the bacteria were widespread in the medium, even a slight disturbance of air inflow and stirring could cause cell death.

In a continuous, semi-continuous, or non-continuous system, the submerged method can manufacture vinegar. Providing fast and effective aeration is the most essential aspect of submerged fermentation systems. The power of the aeration system to break air bubbles allows the flow of oxygen to the bacteria through the medium. It was observed the most significant step in preventing *Acetobacter* cell death. Although conventional open vat methods can be run in a semi-continuous mode and have been utilized in several countries across the world.

6.5.3 APPLICATION

Acetic acid is the main principal in vinegar production and is also used as an essential antimicrobial agent that inhibits the growth of pathogenic and spoilage causing organisms in fermented foods; it also prevents spoilage in beverages such as wine low as 1.2–1.4 g/L concentrations are harmful. Food-grade vinegar, which is used as a preservative and condiment for food worldwide, is a diluted solution of acetic acid formed by acetic acid bacteria. Furthermore, vinegar has been shown to have health-promoting properties. Acetic acid is also used as a solvent in chemical reactions, such as the manufacture of terephthalic acid, and is the primary ingredient for polyethylene terephthalate. Almost 70% of all vinegar manufactured today is supplied to the food industry, where it is used to produce sauces, salad dressings, mayonnaise, tomato products, cheese dressings, mustard, and soft drinks.

KEYWORDS

- **acetic acid**
- **amino acids**
- **Baker's yeast**
- **citric acid**
- **enzymes**
- **L-lysine**
- **microorganisms**
- **submerged fermentation method**

REFERENCES

Angumeenal, A. R., & Venkappayya, D., (2013). An overview of citric acid production. *LWT-Food Science and Technology, 50*(2), 367–370.

Baltz, R. H., Demain, A. L., & Davies, J. E., (2010). *Manual of Industrial Microbiology and Biotechnology*. American Society for Microbiology Press.

Becker, J., Lange, A., Fabarius, J., & Wittmann, C., (2015). Top value platform chemicals: Bio-based production of organic acids. *Current Opinion in Biotechnology, 36*, 168–175.

Bekatorou, A., Psarianos, C., & Koutinas, A. A., (2006). Production of food-grade yeasts. *Food Technology & Biotechnology, 44*(3), 407–415.

Benda, I., & Reed, G., (1982). Production of bakers' yeast. In: Prescott & Dunn's, (eds.), *Industrial Microbiology* (4th edn., p.593.

Bennett, J. W., & Klich, M. A., (1992). *Aspergillus: Biology and Industrial Applications* (pp. 134–7). Boston: Butterworth-Heinemann.

Branduardi, P., Porro, D., & Horst, F., (2012). Yeasts in biotechnology. *Yeast: Molecular and Cell Biology* (pp. 347–370). John Wiley & Sons Inc., Germany.

Burk, M. J., & Van, D. S., (2016). Biotechnology for chemical production: Challenges and opportunities. *Trends in Biotechnology, 34*(3), 187–190.

Canteri-Schemin, M. H., Fertonani, H. C. R., Waszczynskyj, N., & Wosiacki, G., (2005). Extraction of pectin from apple pomace. *Brazilian Archives of Biology and Technology, 48*(2), 259–266.

Cavallo, E., Charreau, H., Cerrutti, P., & Foresti, M. L., (2017). *Yarrowia lipolytica*: A model yeast for citric acid production. *FEMS Yeast Research, 17*(8), fox084, 1–16.

Ciriminna, R., Meneguzzo, F., Delisi, R., & Pagliaro, M., (2017). Citric acid: Emerging applications of key biotechnology industrial product. *Chemistry Central Journal, 11*(1), 1–9.

Crueger, W., & Crueger, A., (1990). *Biotechnology: A Textbook of Industrial Microbiology* (pp. 112–120). Sinaver Associates. Inc., Sunderland.

Eriksson, K. E., (1990). Biotechnology in the pulp and paper industry. *Wood Science and Technology, 24*(1), 79–101.

Friesen, D. T., Babcock, W. C., Brose, D. J., & Chambers, A. R., (1991). Recovery of citric acid from fermentation beer using supported-liquid membranes. *Journal of Membrane Science, 56*(2), 127–141.

Gellissen, G., & Hollenberg, C. P., (1997). Application of yeasts in gene expression studies: A comparison of *Saccharomyces cerevisiae*, *Hansenula polymorpha* and *Kluyveromyces lactis*: A review. *Gene, 190*(1), 87–97.

Glazer, A. N., & Nikaido, H., (2007). *Microbial Biotechnology: Fundamentals of Applied Microbiology*. Cambridge University Press.

Glazer, A. N., Nikaido, H., & Brown, A. J., (1996). Microbial biotechnology: Fundamentals of applied microbiology. *Trends in Biotechnology, 14*(6), 215.

Goldammer, T., (2000). *The Brewer's Handbook–the Complete Book to Brewing Beer.* Chapters 8–12. Apex Publishers, Clifton, Virginia, USA.

González-Sáiz, J. M., Fernández-Torroba, M. A., & Pizarro, C., (1997). Application of weakly basic copolymer polyacrylamide (acrylamide-CO-N, N′-dimethylaminoethyl methacrylate) gels in the recovery of citric acid. *European Polymer Journal, 33*(4), 475–485.

Grewal, H. S., & Kalra, K. L., (1995). Fungal production of citric acid. *Biotechnology Advances, 13*(2), 209–234.

Hossain, M., Brooks, J. D., & Maddox, I. S., (1984). The effect of the sugar source on citric acid production by *Aspergillus Niger*. *Applied Microbiology and Biotechnology, 19*(6), 393–397.

Ikram-ul-Haq, S. K., Ali, S., Ashraf, H., Qadeer, M. A., & Rajoka, M. I., (2001). Mutation of *Aspergillus Niger* for hyperproduction of citric acid from black strap molasses. *World Journal of Microbiology and Biotechnology, 17*(1), 35–37.

Jianlong, W., (2000). Enhancement of citric acid production by *Aspergillus Niger* using n-dodecane as an oxygen-vector. *Process Biochemistry, 35*(10), 1079–1083.

Jianlong, W., Xianghua, W., & Ding, Z., (2000). Production of citric acid from molasses integrated with in-situ product separation by ion-exchange resin adsorption. *Bioresource Technology, 75*(3), 231–234.

Kantar, C., & Honeyman, B. D., (2006). Citric acid enhanced remediation of soils contaminated with uranium by soil flushing and soil washing. *Journal of Environmental Engineering, 132*(2), 247–255.

Kirsop, B. E., & Kurtzman, C. P., (1988). *Living Resources for Biotechnology: Yeasts* (p. 234). UK: Cambridge University Press.

Kolicheski, M. B., (1995). *Production of Citric Acid by Solid-State Fermentation Using Cassava Bagasse as Substrate* (p. 137). MScThesis, Curitiba, Brazil.

Koromyslova, A. D., White, P. A., & Hansman, G. S., (2015). Treatment of norovirus particles with citrate. *Virology, 485*, 199–204.

Kowalski, S., Zander, I., & Windisch, S., (1981). Hybrid yeast strains capable of raising an extraordinarily broad range of dough types. *European Journal of Applied Microbiology and Biotechnology, 11*(3), 146–150.

Kubicek, C. P., Röhr, M., & Rehm, H. J., (1985). Citric acid fermentation. *Critical Reviews in Biotechnology, 3*(4), 331–373.

McMillan, J. D., (2006). *Twenty-Seventh Symposium on Biotechnology for Fuels and Chemicals (No. 129–132).* Springer Science & Business Media.

Mourya, S., & Jauhri, K. S., (2000). Production of citric acid from starch-hydrolysate by *Aspergillus Niger. Microbiological Research, 155*(1), 37–44.

Pallares, J., Rodriguez, S., & Sanroman, A., (1996). Citric acid production in submerged and solid-state culture of *Aspergillus Niger*. *Bioprocess Engineering, 15*(1), 31–33.

Pandey, A., & Soccol, C. R., (1998). Bioconversion of biomass: A case study of ligno-cellulosics bioconversions in solid state fermentation. *Brazilian Archives of Biology and Technology, 41*(4), 379–390.

Pandey, A., Soccol, C. R., Nigam, P., & Soccol, V. T., (2000). Biotechnological potential of agro-industrial residues. I: Sugarcane bagasse. *Bioresource Technology, 74*(1), 69–80.

Papagianni, M., Mattey, M., & Kristiansen, B., (1999). The influence of glucose concentration on citric acid production and morphology of *Aspergillus Niger* in batch and culture. *Enzyme and Microbial Technology, 25*(8, 9), 710–717.

Reddy, N., Reddy, R., & Jiang, Q., (2015). Crosslinking biopolymers for biomedical applications. *Trends in Biotechnology, 33*(6), 362–369.

Ringbom, K., Rothberg, A., & Saxén, B., (1996). Model-based automation of baker's yeast production. *Journal of Biotechnology, 51*(1), 73–82.

Rohr, M., Kubicek, C. P., & Kominek, J., (1983). Citric acid. In: Rehm, H. J., & Reed, G., (eds.), *Biotechnology* (Vol. 3, pp. 419–454). Verlag. Chemmie, New York.

Rößle, C., Gormley, T. R., & Butler, F., (2009). Efficacy of Natureseal® AS1 browning inhibitor in fresh-cut fruit salads applications, with emphasis on apple wedges. *The Journal of Horticultural Science and Biotechnology, 84*(6), 62–67.

Roukas, T., (1998). Carob pod: A new substrate for citric acid production by *Aspergillus Niger. Applied Biochemistry and Biotechnology, 74*(1), 43–53.

Sahm, H., Eggeling, L., Eikmanns, B., & Krämer, R., (1995). Metabolic design in amino acid-producing bacterium *Corynebacterium glutamicum*. *FEMS Microbiology Reviews, 16*(2, 3), 243–252.

Sahoo, B. M., & Banik, B. K., (2019). Baker's yeast-based organocatalysis: Applications in organic synthesis. *Current Organocatalysis, 6*(2), 158–164.

Sarangbin, S., & Watanapokasin, Y., (1999). Yam bean starch: A novel substrate for citric acid production by the protease-negative mutant strain of *Aspergillus Niger*. *Carbohydrate Polymers, 38*(3), 219–224.

Schultz, A. S., & Atkin, L., (1947). The utility of bios response in yeast classification and nomenclature. *Archives of Biochemistry and Biophysics, 14*, 369–380.

Schultz, A., Atkin, L., & Frey, C. N., (1937). A fermentation test for vitamin B1. *Journal of the American Chemical Society, 59*(5), 948–949.

Seligra, P. G., Jaramillo, C. M., Famá, L., & Goyanes, S., (2016). Biodegradable and non-retrogradable eco-films based on starch–glycerol with citric acid as crosslinking agent. *Carbohydrate Polymers, 138*, 66–74.

Soccol, C. R., & Vandenberghe, L. P., (2003). Overview of applied solid-state fermentation in Brazil. *Biochemical Engineering Journal, 13*(2, 3), 205–218.

Soccol, C. R., Vandenberghe, L. P. S., & Lebeault, J. M., (1999). *Production of Citric Acid by Solid-State Fermentation Using A. Niger*. Patent Pt. Br. DEINPI/PR, 175.

Soccol, C. R., Vandenberghe, L. P., Rodrigues, C., & Pandey, A., (2006). New perspectives for citric acid production and application. *Food Technology & Biotechnology, 44*(2), 141–149.

Soccol, C. R., Vandenberghe, L. P., Rodrigues, C., Medeiros, A. B. P., Larroche, C., & Pandey, A., (2008). Production of organic acids by solid-state fermentation. In: *Current Developments in Solid-State Fermentation* (pp. 205–229). Springer, New York, NY.

Sun, D., Chen, Y., Tran, R. T., Xu, S., Xie, D., Jia, C., & Bai, X., (2014). Citric acid-based hydroxyapatite composite scaffolds enhance calvarial regeneration. *Scientific Reports, 4*(1), 1–9.
Trivedi, N. B., Cooper, E. J., & Bruinsma, B. L., (1984). Development and applications of quick-rising yeast. *Food Technology (USA), 38*(6), 51.
Trivedi, N. B., Jacobson, G. K., Tesch, W., & Friend, J. P., (1986). Bakers' yeast. *Critical Reviews in Biotechnology, 4*(1), 75–109.
Umemura, K., Ueda, T., Munawar, S. S., & Kawai, S., (2012). Application of citric acid as a natural adhesive for wood. *Journal of Applied Polymer Science, 123*(4), 1991–1996.
Vandenberghe, L. P. S., Pandey, A., Soccol, C. R., & Lebeault, J. M., (1999). *Citric Acid Production by Aspergillus Niger*, 26–29.
Vandenberghe, L. P., Soccol, C. R., Pandey, A., & Lebeault, J. M., (1999). Microbial production of citric acid. *Brazilian Archives of Biology and Technology, 42*(3), 263–276.
Vandenberghe, L. P., Soccol, C. R., Pandey, A., & Lebeault, J. M., (2000). Solid-state fermentation for the synthesis of citric acid by *Aspergillus Niger*. *Bioresource Technology, 74*(2), 175–178.
Vandenberghe, L. P., Soccol, C. R., Prado, F. C., & Pandey, A., (2004). Comparison of citric acid production by solid-state fermentation in flask, column, tray, and drum bioreactors. *Applied Biochemistry and Biotechnology, 118*(1), 293–303.
Vanderberghe, L., Pandey, A., Lebeault, J., & Soccol, C., (1999). Producao de ácido cítrico por/*Aspergillus Niger*/LPB 21 em fermentacao no estado sólido com casca de café. *SEMINARIO Internacional sobre Biotecnologia na Agroindústria Cafeeira, 3.* Londrina (Brasil), Anais.
Vergano, M. F., Soria, M. A., & Kerber, N. L., (1996). Influence of inoculum preparation on citric acid production by *Aspergillus Niger*. *World Journal of Microbiology and Biotechnology, 12*(6), 655–656.
Waites, M. J., Morgan, N. L., Rockey, J. S., & Higton, G., (2009). *Industrial Microbiology: An Introduction.* John Wiley & Sons.
Ward, O. P., (1995). Application of baker's yeast in bioorganic synthesis. *Canadian Journal of Botany, 73*(S1), 1043–1048.
Xu, D. B., Madrid, C. P., Röhr, M., & Kubicek, C. P., (1989). The influence of type and concentration of the carbon source on production of citric acid by *Aspergillus Niger*. *Applied Microbiology and Biotechnology, 30*(6), 553–558.
Yin, X., Li, J., Shin, H. D., Du, G., Liu, L., & Chen, J., (2015). Metabolic engineering in the biotechnological production of organic acids in the tricarboxylic acid cycle of microorganisms: Advances and prospects. *Biotechnology Advances, 33*(6), 830–841.
Yokoya, F., (1992). Citric acid production. In: *Industrial Fermentation Series* (pp. 1–82). Campinas, SP, Brazil.
Yokoya, F., (1992). *Fermentação cítrica, Fundação Tropical de Pesquisas e Tecnologia "André Tosello* (pp. 1–82). Campinas, São Paulo, Brazil.
Yoshida, A., & Yamataka, A., (1953). On the metaphosphate of yeast. I. *The Journal of Biochemistry, 40*(2), 85–94.

CHAPTER 7

Down Stream Processing

7.1 INTRODUCTION

After producing the product, extraction and purification are the most significant steps in any biological process. The steps necessary are incredibly time-consuming and costly. As a result, downstream biological processes must be efficient and cost-effective for the strategy to be economically viable. However, recovery costs in microbiological methods typically range between 15% and 75% of the entire manufacturing cost. The recovery price is determined by various factors, such as the nature and degree of purity of the product and the method employed in the recovery process.

On the other hand, downstream processes are influenced by production characteristics such as pH, temperature, plant handling, process maintenance, and aseptic conditions. It is well established that extracellular and straightforward products have a lower production cost; intracellular goods, on the other hand, have a higher recovery cost. As a result, the following criteria are considered when developing a process for recovery of the product:

- Intracellular/extracellular location of that specific product.
- Product concentration in the fermented broth.
- The physical and chemical characteristics of the product may be used to aid in selecting the separation technique to be used.
- Required degree of purity of that particular product as determined by the appropriate authority (like FDA).
- Product price in reference to market competitor for same product.

Along with all of these elements, bio-safety considerations in the manufacturing process must be taken into account. The removal of cellular fragments and medium components is the initiation phase in recovering an extracellular product. Centrifugation is the predominant technique used to separate

Handbook of Industrial Food Microbiology. Manju Nehra & Vikash Nain
© 2025 Apple Academic Press, Inc. Co-published with CRC Press (Taylor & Francis)

cellular components. The following procedures are used to fractionate the broth and extract the product. Ultrafiltration technique, reverse osmosis methods, adsorption technology, ion exchange technique, gel filtration or affinity chromatography, liquid-liquid extraction, two-phase aqueous extraction, or precipitation technologies may be utilized for the isolation or purification of the desired final product. The method of choice is determined by the nature of the product being produced. Following that, the product is crystallized in order to achieve the desired purity. The modification of this system may allow for the isolation of by-products. If you take care of the following processes in the upstream or bioconversion process, you may find that product recovery is less difficult.

- By adopting microorganisms that do not release any pigment during the fermentation, including those that do not generate any undesirable metabolites.
- By altering the production conditions in such a way that the least amount of unwanted secondary metabolites is produced.

The following process variables should be thoroughly investigated and meticulously maintained:

- Harvesting season;
- Maintaining a constant pH level during fermentation and harvesting;
- Temperature monitoring and control;
- Addition of flocculation and separation reagents harvesting time.

There are a variety of various routes that can be used to purify a large number of different chemicals. Capital cost, processing cost, throughput need, yield potential, product quality, and technical skill are all factors which are to be critically considered while determining the best process for a given application at the end.

7.2 MICROBIAL CELLS AND OTHER SOLID MATTER REMOVAL

Centrifugation or filtration separates the microbial cells and other solid particles from the broth during the fermentation process. Heat treatment or the addition of flocculating agents may be necessary in some circumstances to boost the sedimentation rate. Recently, some modern techniques for separating solid materials from fermented broth, such as electrophoresis and dielectrophoresis, have been investigated to take advantage of the charge properties of the microbial cells to separate the solid materials from the fermented broth. The other procedures include ultrasonic treatment, which is used to improve the flocculating qualities of the product, and the use of magnetic field. On the other hand, these techniques are burdened by high costs and scaling issues. The two-phase liquid extraction process has gotten a lot of attention lately because of its ease of scaling up and cost-effectiveness and the fact that it may be utilized for materials that require gentle handling.

7.2.1 FOAM SEPARATION

It is primarily determined by the surface features of the materials that foam separation of solid materials such as microbial cells or proteins and colloidal components occurs. Cells or molecules are absorbed on the surfaces of gas bubbles that form as a result of the bowling air being blown through them, and they are then separated by skimming. Surfactants that are employed in general are referred to as collectors, and the materials that are made surface-active and collected are referred to as colligends. Surfactants are utilized in general to collect and surface-activate materials. It is mandatory to consider the following factors while developing a process on the basis of this technique: pH, temperature, airflow rate, and the ratio of surfactant to colligend collection (if applicable). Figure 7.1 depicts a flow chart representation of the foam separation process.

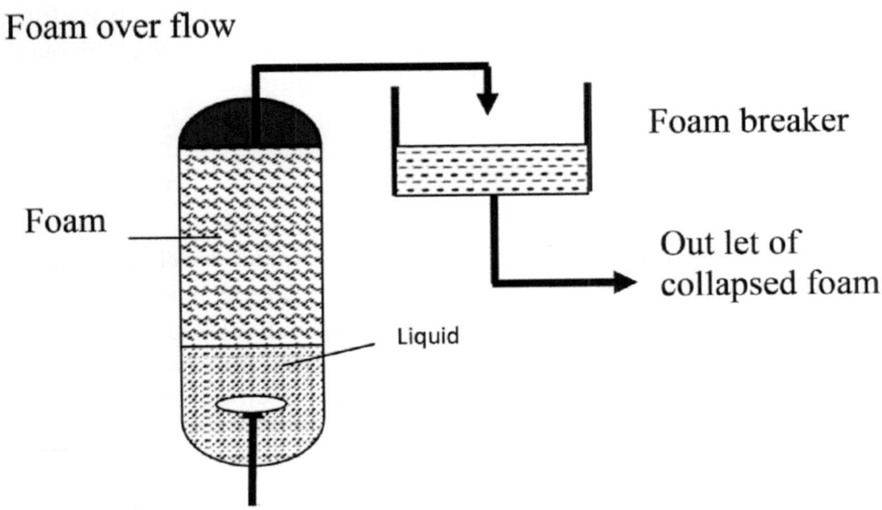

FIGURE 7.1 Schematic representation of the foam separation process.

7.2.2 PRECIPITATION

Generally speaking, precipitation is an aftermarket product recovery method that can be applied at any level of the downstream processing chain. This method helps separate some goods in one phase and removing specific contaminants or impurities in the other. In some cases, the method enables the enrichment and concentration of the product while also reducing the volume of material handled. Precipitation of various materials is accomplished through the application of the procedures listed below:

- To arrive at the isoelectric point of the chemical, the pH of the solution must be changed. As the solubility of the molecules decreases, they start to precipitate.
- The most commonly employed salts (sulfates) to precipitate proteins are ammonium and sodium salts. Salinity removes water molecules off the surface of protein molecules, enabling the molecules' non-polar regions to be more easily exposed to the environment. Because of the non-polar interaction of the molecules, they aggregate and precipitate in the presence of water.
- Solvents (such as cold ethanol or acetone) are employed to precipitate proteins from the broth by altering the dielectric characteristics of the molecules in the broth. Similar to ionic polymers, non-ionic polymers such as polyethylene glycol precipitate proteins similarly.
- Protein aggregation and precipitation are also accomplished using polyelectrolytes. Several of the more recent methods, such as affinity precipitation, which is the selective precipitation of compounds, protein precipitation, dye binding, and aggregation, are particularly intriguing.

7.2.3 FILTRATION

With various filtration methods, filtration is the most widely utilized and efficient procedures for solid-liquid separation. The technique employs a porous media that enables the passage of gases and liquids but not of solids. It is scalable. However, the following considerations influence the selection of the most appropriate equipment to match the specific requirement while remaining economically viable.

- Filtrate density and viscosity;
- Nature of solid particles such as shape, size, distribution, and packing characteristics;

- Solid: liquid ratio;
- Scale of operation;
- Material required to be recovered, i.e., liquid/solid or both;
- Type of operation: batch or continuous;
- The requirement for an additional vacuum suction attachment or the requirement for low temperature.

7.2.3.1 THEORIES OF FILTRATION

It is possible to create inexpensive filtration equipment by using filtration fabric and porous material to support the cloth. The solid material is a deposition and the form a cake during the passage of broth through the filter cloth. In response to the slow deposition on filter, the cake thickens and a gradual buildup of resistance to flow occurs. To maintain a consistent flow rate, an increasing amount of pressure must be applied to the cloth in proportion to the rise in the thickness of the cake and the resistance to flow. When particles are compressed, the apertures of the clarifying cloth are closed, and the flow is stopped. This can happen when the pores of the filtering cloth are blocked due to clogging. For the filtration process to be effective, pressure must not be applied, especially when the particles are compressible (Table 7.1; Figures 7.2 and 7.3).

Darcy's equation can be used to represent the flow through a bed of constant and uniform depth:

$$\text{Flow rate} = dV/dt = KA\Delta P/\mu L \quad (1)$$

where μ = liquid viscosity; L = depth of the filter bed; ΔP = the difference in the pressure across the filtration bed; A = area of the filtration bed exposed to the liquid to be filtered; K = the constant for the filtration bed (known as Kozeny's constant); *K* depends on the specific surface area *s* (specific surface area is defined as the surface area/unit volume) of the particles that make up the filtration bed and the voidage (Σ) when the particles are packed together. The voidage is defined as the fractional volume of the filter bed that is available for the liquid to pass through and it is generally 0.3 to 0.6. Therefore, *K* can be expressed as:

$$K = \Sigma/5(L - \Sigma)^2 s^2 \quad (1)$$

TABLE 7.1 Different Types of Filtration Systems

Mode of Filtration	Types of Filters	Characteristic Features and Applications
Batch filtration	**Plate and Frame Filter:** It is one type of pressure filter consisting of plates and frames arranged alternately; frames are fitted with filtering elements (filter pads or filter cloth). The assembly of filters and frames are held together by a horizontal structure and forms a series of liquid tight components. Liquid passes through the body through a hole in the corner, flows through the grooves and discharged through an outlet tap.	In industrial scale this is the cheapest filtration process and the area required is less. The main drawback of this process is the frequent wear and tear of the filtering cloth in batch operations due to its systematic dismantling. This system finds wide application for clarification in the breweries. It is also used for collecting high value solids for which the application of a continuous filtration apparatus is not suitable.
	Pressure Leaf Filter: These filters consist of a number of metal plates called leaves with metal framework of grooved plates. These are covered with fine wire mesh or sometimes a filter cloth, occasionally are pre-coated with a cellulose fiber coat. The slurry for the filtration is fed into the filter by pressure or by suction. The system can be steam sterilized. Types: • Vertical metal-leaf filter. • Horizontal metal-leaf filter. • Stacked-leaf filter.	Generally used for clarification of high-volume liquid containing low solid load or in small batch filtration of valuable solids.
Continuous filtration	**Rotary Vacuum Filter:** The filter consists of a hollow rotating drum. The drum is segmented internally. The drum remains covered with a piece of fabric or metal filter and remains immersed in a trough containing the liquid to be filtered. Vacuum is applied inside the drum and the slurry is fed into the container from outside. Due to the vacuum, filtrate is sucked into the drum and through the compartments collected in a collecting vessel. To help the discharge of the filter cake air pressure is applied inside the drum and remaining broth is carefully washed out from the cake by water jet without allowing much dilution of the filtering liquid (broth). Depending on the cake discharge pattern, a number of rotary vacuum filters are available. These are: • String discharge. • Scraper discharge. • Scraper discharge with precoating of the drum.	This is applied in the industries for the filtration of large volume of liquids in a continuous process. Though the process can be used for the separation of calls from the broth, but disrupted cell slurry can also be effectively removed from the processing liquids.

TABLE 7.1 Different

Mode of Filtration	Types of Filters	Characteristic Features and Applications
Cross flow filtration (tangential flow filtration)	Cross flow filtration system contains a media storage tank, a pump and a method of packs of membrane. To attain a much-convoluted surface and an increased surface area the membrane is usually packed in a cassette of hollow fibers or flat sheet in a plate and frame type stack or a spiral cartridge. The broth is forced across the membrane surface, most of the bulk material sweeps the membrane and returns to the main tank and less than 10% pass through the membrane as permeate. As the process continues, the volume of the original feed reduces to 5–10% of their initial volume. [Two types of membranes can be used depending on the need of the process: microporous membrane with a specific pore size (0.45 μM or 0.22 μM) or an ultrafiltration membrane with a specific molecular weight cut-off].	The process is highly efficient and >99.9% cell retention can be achieved. The system is a closed system, and no aerosol is formed, the separation process is independent of media and cell densities and no addition of filter aid is needed. The system can be sterilized with slight modification. It is used for the filtration of fermentation broths, cell lysates, etc. *Note:* Factors that influence crossflow filtration. Increased pressure drop increases the rate of filtration up to a specific limit. But as the system is based on the swept clean principle pressure drop should not be too high else it will lead to the blockage of the membrane. Therefore, to achieve the best results higher flow rate is recommended.
–	–	Higher temperature increases the flow rate by lowering the viscosity of the system.
–	–	The membrane may get clogged by the fowl material present in the medium. This can be controlled by modifying the constituents of membrane or modifying the media, like, minimizing the addition of antifoam reagent.

where • and *s* cannot be determined easily and in most practical cases *L* also cannot be measured directly. *L* can be expressed as:

$$L- = vV/A$$

where V = the broth volume passed in time t; v = volume of the cake deposited per unit volume of the brothAfter substitution in equation (1).

$$dV/dt = KA^2\Delta/\mu vV \quad (2)$$

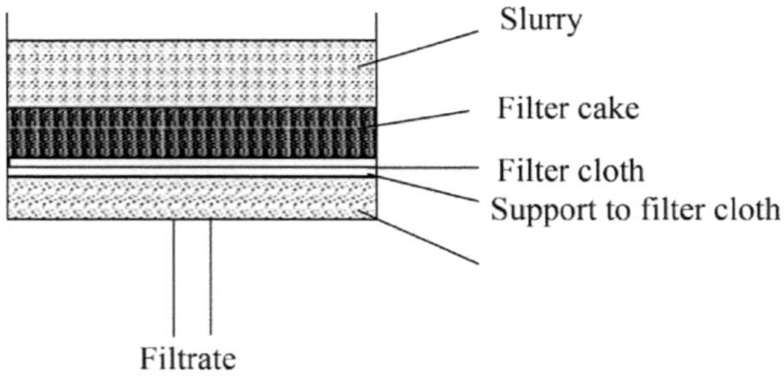

FIGURE 7.2 Schematic presentation of a simple filtration apparatus.

Eqn. (2) is a generalized equation that relates the filtration rate to the pressure drop, the filter's cross-sectional area, and the filtrate retained. Filtration at constant pressure can be described using the integrated equation.

$$\mathrm{VdV} = KA^2\Delta P\, \mathrm{d}t/\mu\ v \tag{3}$$

Integrating Eqn. (3):

$$V^2 = 2KA^2\ \Delta \mathrm{P}t/\mu v \tag{4}$$

In Eq. (4) ΔP is a constant and μ is generally equal to 1 and v can be determined in the laboratory, whereas, $A2$ remains approximately constant. Therefore, it can be inferred that there is a linear relationship between $V2$ and t. It is generally possible to calculate the pressure necessary to maintain a constant filtration rate, but is of little practical use.

7.2.3.2 USE OF FILTER AIDS

In most cases, filter aid is required when the filtering material clogs the filter media (filter cloth) due to the presence of bacterial broth or gelatinous material in the filtering material. Diatomaceous earth is the most often encountered filtering aid (kieselguhr). The voidage is extremely high (about 0.85), and the porosity rises when mixed with the cell suspension. To reduce the absorption of the filter material, the filter aid should be used sparingly. It can be used as an early bridging material in a filter medium's larger pores to prevent or reduce pore blindness, which can be harmful. Application of a thin coating of the filter aid to the filter medium before the broth's filtration is necessary and it build up a suitable filter bed before beginning the filtering process to ensure an efficient filtering process when filter aids are used. It is blended with the harvested broth after a proper amount of filter bed has been added to it. Following bed formation, the raffinate is introduced again into the residual soup to commence the filtration process, whereas a thick precoat filter is built-up on the drum of a vacuum drum filter fitted with an advancing knife. Filter aids are used to facilitate in the purification of isolated intracellular products in general. Filter aids are not an option when it comes to generating microbial biomass or cells that require pretreatment treatments such as flocculation or heating.

7.2.4 APPLICATIONS

Tangential flow filtration, also known as Cross flow filtration, is a method of filtration in which the filtered solution is passed over the membrane's surface without undergoing any substantial phase changes. Because of its intrinsic resistance to fouling, crossflow filtration is virtually exclusively used in the industrial setting. Any particles accumulated on the membrane's surface are eliminated by the shearing force caused by the fluid moving across it. This helps to keep the flux across the membrane reasonably constant. In commercial food industries such as fermentation, food ingredients, dairy, brewery, beverage, nutraceutical, and starch and sweetener, etc., crossflow membrane filtering is quickly gaining global acceptance as a critical production step in a variety of process lines. This filtering technique is used extensively in the food sector to process various beverages, alcoholize beer and wine, reclaim water, and clarify corn syrup and fructose. It can be used to concentrate egg whites or whole eggs. Membrane filtration's ability to create precise separations at low or ambient temperatures without phase change can make it a far more cost-effective procedure than various traditional methods like rotary vacuum filtration or filter presses.

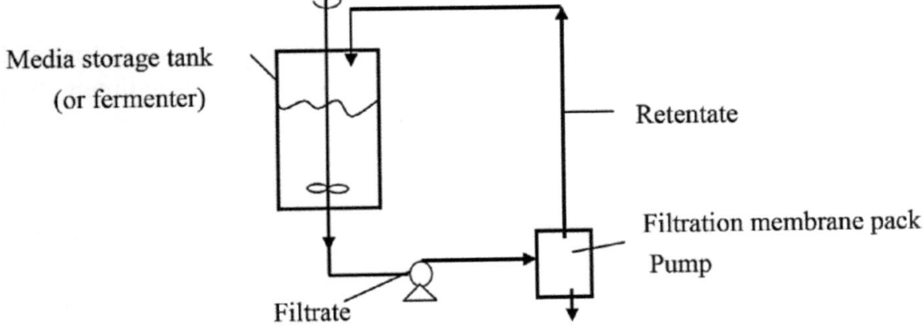

FIGURE 7.3 Schematic presentation of a typical crossflow filtration system.

7.2.5 CENTRIFUGATION

When filtering fails to provide sufficient solid separation or when a particularly greater standard of segregation is desired, centrifugation is used. The concept of settling under centrifugal force underlies the centrifugation process.

The sedimentation rate of a spherical solid suspended in a liquid with Newtonian viscosity is proportional to the square of the particle diameter, according to the Strokes law.

As a result, a particle's rate of sedimentation under gravitational force can be stated as:

$$\mathrm{Vg} = d^2 g\,(\rho_P - \rho_L)/18\mu \qquad (5)$$

where Vg = rate of sedimentation; d is the particle diameter; g is the gravitational constant; ρ_P is the density of the particle; ρ_L is the density of the liquid; and μ is the viscosity.

The equation can be modified for sedimentation under the centrifugal force as:

$$\mathrm{Vc} = d^2 \omega^2 r\,(\rho_P - \rho_L)/18\mu \qquad (6)$$

where Vc = rate of sedimentation under centrifugation force; ω = angular velocity of the rotor; r = radial position of the particle.
Taking Eqn. (6) and dividing it by Eqn. (5), we get:

$$Vc/Vg = \omega\acute{\omega}^2 r/g \qquad (7)$$

The equation is a measure of centrifugal force's sedimentation power in comparison to gravitational force. It is symbolized by the sign Z and named as relative centrifugal force (RCF).

As indicated above, other parameters such as the viscosity differential between the cell mass and the liquid play a modest impact on sedimentation rate. While it is true that when the temperature rises, the viscosity of the liquid varies, the effect can be ignored because the viscosity difference between the particles and the liquid is substantially less. Thus, while optimizing sedimentation via centrifugation, the rotor diameter and angular velocity of the rotor are critical elements to consider. Generally, batch centrifugation with tiny volumes can also be utilized in the laboratory, but they are not practicable in industry. Centrifuges are being used in enterprises in a continuous or semi-continuous manner.

7.2.5.1 DIFFERENT TYPES OF CENTRIFUGES

For diverse industrial uses, multiple types of centrifuges with various dimensions and modalities of operation are available. They differ mainly in their operation, capacities, speed, loading methods, and discharging. The type of centrifuge to use is determined by the application and efficacy of centrifuges.

Table 7.2 describes different types of centrifuges used in the industries (excluding ultracentrifuge and another advanced centrifugation used for the particular type of application).

TABLE 7.2 Different Types of Centrifuges

Types of Centrifuges	Description	Application
Basket centrifuge (perforated-bowl basket centrifuge)	Contains perforated bowls with a filter bag of nylon or cotton. Liquid is fed in continuous mode, and it is operated at not more than 4,000 rpm. Solid biological mass can be washed before removing the bowls. Rate of feeding is 50–300 dm^3/min and have a solid holding capacity of maximum 500 dm^3.	Used for the separation of mycelial mass, molds, and crystalline compounds.
Tubular-bowl centrifuge	The main component of this system is a cylindrical bowl which may be of variable design depending on their application. The system remains suspended by a shaft which is flexible and rotates by a motor fitted overhead. The inlet is at the bottom fitted with a nozzle which enters through the bottom bearing. The inlet consists of a solid and a liquid phase; the liquid generally contains a light and a heavy phase.	The system is applied for the separation of particles having dimension of 0.1 μm to 200 μm in diameter and for a liquid having solid load of only 10% or less. Can be applied for – (i) light phase/heavy phase liquid separation; (ii) for the separation of solid/light phase liquid/heavy liquid phase and only for; and (iii) solid/liquid separation.
	The solid particles sediment on the rotor wall and the two liquid phases gets separated into two distinct zones. At the exit, they are kept separated by the ring's adjustable lips, which may be of various sizes.	This system can be operated at a speed of 50,000 rpm. Limitations are limited solid holding capacity, loss of efficiency, etc.

TABLE 7.2 *(Continued)*

Types of Centrifuges	Description	Application
Solid bowl scroll centrifuge (decanter centrifuge)	The central part of this system is a horizontal rotating solid bowl. Slurry is fed to the bowl through the spindle of an Archimedean screw within the bowl. Solid is separated on the bowl wall and is scraped up to the conical end of the bowl. The slope of the bowl is so set that the excessive liquid is drained out from the solid. Fluid is discharged at the other end of the bowl.	The system is used in continuous mode for the separation of solids from the fermentation broth. The bigger version can be applied at a speed of maximum 5,000 rpm, smallest versions can be applied sometimes at a speed of 10,000 rpm. The system is available for various applications: • with the facility of cake washing; • with vertical bowl decanting facility; • with the facility of in-situ cleaning; and • the facilities for the containment of biohazards.
Multichamber centrifuge	This system consists of multiple chambers mounted within the chamber of the rotor. The slurry is fed into the chambers through a system of spindles and travels through the chambers through a circuitous route. At the outer face of each chamber the solid is collected.	The system is applied to separate particles with a dimension of 0.1 μm to 200 μm in diameter. Though the system is having greater solid handling capacity and there is virtually no loss of efficiency, but its mechanical stability limits its larger application and speed maxima (6,500 rpm).
Disc-bowl centrifuge	This system consists of a central inlet pipe and a system of conical disc, made-up of stainless steel arranged in stacks with a spacer. The broth to be separated is injected into the system through a central pipe, and then it flows outward toward the discs. It flows upward and inward in between the disc at an angle 45° with the axis of rotation. Within the closed system the solid sediments rapidly, accumulating on the inner wall of the bowl. The sediment formed is not slid in this case, forms slurry which flows and can be discharged continuously.	The system is highly efficient, has capacity of high-volume liquid handling, easy removal of solid and *in-situ* cleaning facility. Solid can be removed by automatic opening of the solid collection bowl. Feed rate is as high as 1,800 dm^3/min and rotational speed ranges from 5,000 rpm to 10,000 rpm.

7.2.6 APPLICATION

To clarify wine, remove particles from fermented broths, etc., centrifugation is preferred. One of the key advantages of centrifugation is its rapidity of clarification. It is useful when early bottling is needed. Centrifugation is also quite beneficial for highly turbid wines. They commonly emit disagreeable scents during spontaneous clarification and frequently clog filter units. Additionally, centrifugation improves the efficiency of polishing filtering by eliminating the majority of suspended particles.

7.2.7 CELL DISRUPTION

The stiff cell wall protects microbes from the external environment. The cell wall can be difficult to break down, and recovering intracellular products necessitates breaking down the cell wall. Numerous cell disintegration methods are available (Figure 7.4), but the method's choice depends on its suitability for the particular substance. Even though, in some circumstances, a specific product can be retrieved from cells with increased yield and purity by employing a specific technology, these types of techniques are not viable. On the other hand, a complete rupture of the cell releases many products, which makes downstream processing more challenging. Sometimes, implementing a certain technique is only viable at the laboratory scale. As a result, selecting suitable procedures for industrial application is a question of research. In recent years, enzymes have emerged as the most important intercellular products of particular interest. However, the disruption methods used for the liberation of enzymes must maintain their activity and proper folding. The yield of enzymes released must be substantial at the precise release time. Some cell disruption techniques are shown in Figure 7.4.

7.2.8 CHEMICAL METHODS

7.2.8.1 DETERGENT TREATMENT

Detergents cause cell wall destruction by interfering with the lipoproteins present in the cell membrane of the microbes and causing the intracellular enzyme to be released into the environment. Various detergents are utilized, including anionic, cationic, and non-ionic. Some of the generally used detergents are SDS (sodium dodecyl sulfate), Quaternary ammonium salts,

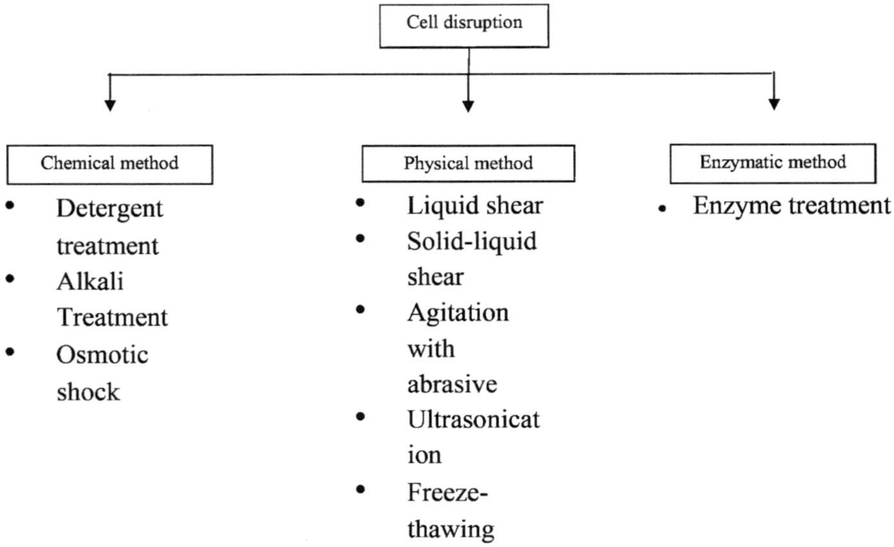

FIGURE 7.4 Different methods for the cell disruption.

Triton X-100, etc. Although these chemicals can stimulate the release of an enzyme, it is important to remember that they can cause protein denaturation, which must be avoided by removing them from the cell-free extract as soon as possible during the subsequent purification process. Application of Triton X-100 is well-known for its ability to liberate enzymes that are bound to membranes. Triton X-100 and guanidine–HCl are two very useful solutions for freeing various proteins from their binding sites.

7.2.8.2 ALKALI TREATMENT

The method used for the cell disruption in minimal cases is treatment with alkali. When an enzyme is very alkaline resistant and can endure a pH of at least 11.4, it is employed. The release of L-asparaginase is among the very few examples of its applications in enzyme release.

7.2.9 PHYSICAL METHODS

7.2.9.1 LIQUID SHEAR

It is most commonly employed for the large-scale release of enzymes or intracellular products from microbial cells that liquid shear stress is applied.

Because of the significant pressure drop in the cell slurry, the shear force exerted through cavitation in the cell slurry is utilized in this method. The machine is composed of a hollow cylinder made of stainless steel and a piston with a system of adjustable valves that is adequate for the application. The cylinder is loaded with cell slurry, and the pressure inside the cylinder is boosted to thousands of pounds per square inch (psi) through a hydraulic system. When a small opening is permitted to pass through the cell slurry, the cells experience an abrupt reduction in pressure. Cavitation occurs due to the pressure drop, and the shock waves that are generated rupture the cells. The pressure decrease that the cells are subjected to directly impacts the disruption of the enzyme release that occurs from the cells. So, it can be deduced that a more significant increase in pressure drop is much more effective at disrupting cell division. Microorganisms with an excellent mechanical property, as well as their form and size, play a crucial role in creating successful disruption. To prevent heat conformational changes of the enzyme during the cell disruption process, an appropriate cooling system will always be there during the procedure. The entire operation takes place between 0°C and 4°C. It is necessary to run the procedure in multi-pass mode for longer to make it less abrasive. To achieve cost-effectiveness in an industrial process, it is necessary to strike a healthy balance between the entire release caused by successful breakdown of the cells and the percentage of enzyme released.

7.2.9.2 SOLID-LIQUID SHEAR

This approach is suitable for use in the laboratory, and it can also be used in the industrial setting on a modest scale in a semi-continuous mode. Frozen specimens of microorganisms are pushed through a tiny hole at high pressure while maintaining a shallow temperature (–25°C). The breakage happens due to the shear produced by the liquid and the existence of ice crystals. The procedure is beneficial for enzymes that are sensitive to changes in temperature.

7.2.9.3 AGITATION WITH ABRASIVES

Cells are disrupted in this procedure with the assistance of physically resistant beads consisting of glass, alumina, or titanium compounds. Cells are stirred with beads inside a hollow compartment by an agitator shaft system housed within the chamber. Cell disintegration is caused by the shear forces generated due to this process. The procedure is exothermic, and a reliable cooling mechanism must be used to prevent the enzyme from becoming saturated due to the heat generated. After a thorough examination, it is possible to apply the procedure broadly.

7.2.9.4 ULTRA-SONICATION

To break the cells on a small scale, ultrasonic pulses are applied. The cavitation that happens as a result of the operation is caused by sonic waves and shear force. A large amount of electrical power is turned into mechanical energy in the form of a sonic wave, which proliferates through a horn in a liquid and causes cavitation to occur. There are various disadvantages to the system, including excessive power usage, heat dissipation, a tiny operational volume, etc.

7.2.9.5 FREEZE-THAWING

It is a really gentle procedure. The formation of ice crystals, the formation of apertures in the cell wall, and the facilitation of enzyme release are all caused by the frequent freezing and thawing of microbes. The approach can be used alone or in conjunction with other techniques.

7.2.9.6 OSMOTIC SHOCK

The application of osmotic shock is used to cause a moderate release of enzymes from the cells. It is interrupted when there is a sudden shift in the salt content within the cells because the osmotic equilibrium is altered within the cells. However, the approach is not very effective for microbiological cells with a strong cell wall, such as bacteria. For this approach to be effective against microbial cells, the cell wall must first be rendered weak in another way before it can be used. In releasing the luciferase enzyme from *Photobacterium fischeri*, the approach has proven to be extremely effective and one-of-a-kind.

7.2.10 ENZYMATIC METHOD

Various enzymes can break different types of bonds found in the cell wall, hence facilitating the liberation of enzymes. The procedure is the gentlest method of releasing intracellular enzymes currently available. Enzymes isolated from *Streptomyces* sp., *Penicillium* sp., *Trichoderme* sp., and snails are also used in the procedure, in addition to lysozyme. The process is quite expensive, but it is extremely successful for releasing enzymes that contain fewer contaminants. It is necessary to eliminate the enzyme that was employed during the downstream step. To attain the greatest results, this method is sometimes combined with other methods of cell disintegration.

7.3 LIQUID CHROMATOGRAPHY

Chromatography is the study that examines the separation of molecules based on differences in their form, size, structure, and/or composition. It is a branch of chemistry that is a subfield of physics. The term "sorption-desorption" refers to a procedure in which sorption-desorption stages on the stationary phase separate sample components carried by the liquid or gaseous phase. Characteristics with more positive interactions with the support will move more slowly than those with negative interactions. It is possible to segregate various pieces as they move across the support material. Various mobile and stationary phases are utilized in chromatographic separations, including immobilized silica gel on glass plates (thin layer chromatography), volatile gases (gas chromatography), paper (paper chromatography). In liquid chromatography, liquids containing hydrophilic, insoluble molecules (hydrophilic insoluble molecules). Liquid chromatography is among the most widely used techniques because it is incredibly versatile and can separate various compounds ranging from lower organic molecules to high importance biodrugs such as enzymes, nucleic acids, carbohydrates, fats, vitamins, and other substances.

There are different modes of liquid chromatography, which are explained in subsections.

7.3.1 ION-EXCHANGE CHROMATOGRAPHY

7.3.1.1 PRINCIPLE

Ion-exchange chromatography is based on the attraction of molecules with opposing charges. The majority of ion-exchange separations take place in columns packed with an ion-exchanger. Ion exchangers are categorized in two classes: cationic exchangers and anionic exchangers. The stationary phase has a negative charge, whereas the mobile phase has a positive counter ion, and anion exchangers have a positive charge in the stationary phase. In comparison, an anion exchanger has a positive-charged fixed phase and a negative-charged mobile phase. This method effectively separates naturally occurring acidic, basic, or neutral compounds. In this type of separation, matrix-bound charged groups reversibly adsorb oppositely charged sample molecules. Diffusion of the molecule via the exchanger matrix is dependent on the degree of cross-linking of the exchanger and the buffer's ionic strength. This diffusion process is capable of regulating the entire ion-exchange reaction's rate. Desorption of the bonded ion, on the other hand, can be performed by altering the pH or salt concentration or by adding competing counter ions, as this tends to break the matrix's electrostatic link with the bound molecule. It is the most often used technology for sorting proteins according to their net charges (Figure 7.5).

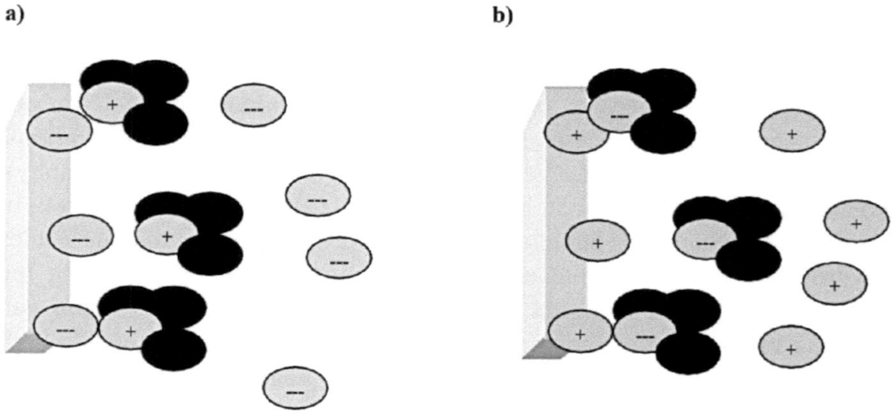

FIGURE 7.5 (a) The stationary phase of a cation exchanger has a negatively charged functional group that attracts positively charged cations. And (b) Anion exchanger: this type of exchanger preserves negatively charged anions by exhibiting a positively charged functional group.

7.3.2 ION EXCHANGE CHROMATOGRAPHY'S COMPONENTS

A typical ion exchange column is composed of ion-exchange resins bonded to inert polymeric particles with a small diameter, with the resins being packed into the column. In the field of ion exchange chromatography, a wide range of matrices and ionic groups can be used to achieve excellent results. Polystyrene, cellulose, and agarose are the most commonly utilized matrices. In addition to sulfonic acid and carboxylic acid, quaternary ammonium groups are also functional ionic groups. Diethylaminoethyl cellulose (DEAE-C) acts as the supporting material, any derivative of amino group having positive charge, as seen in Figures 7.6(a) and 6(b). On the other hand, a cation exchanger may be composed of carboxymethyl cellulose (CMC)-based support material that has been derivatized with a negatively charged carboxylic acid. Table 7.3 shows several examples of ion-exchangers in practical applications.

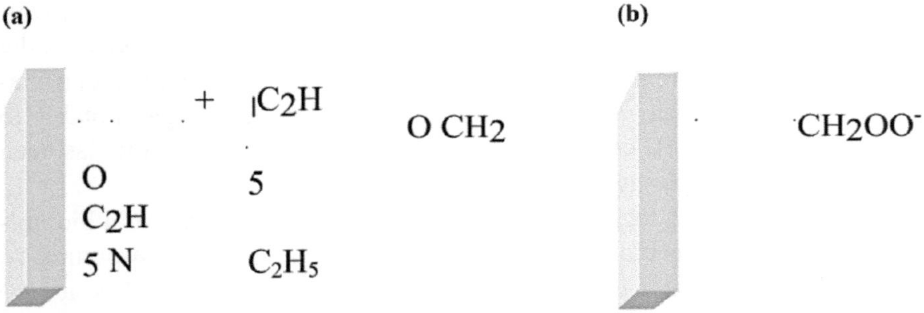

FIGURE 7.6 (a) Diethylaminoethyl cellulose based support material; and (b) CM-cellulose based support material.

TABLE 7.3 Examples of Commonly Used Ion-Exchangers

Anion Exchanger	Functional Groups
Quaternary aminoethyl (QAE)	$–O–CH_2CH_2N^+(C_2H_5)_2–CH_2–CHOH–CH_3$
Quaternary ammonium (QA)	$–O–CH_2–CHOH–CH_2–O–CH_2–CHOH–CH_2–N^+(CH_3)_3$
Cation Exchanger	Functional Groups
Sulfopropyl (SP)	$–O–CH_2–CHOH–CH_2–O–CH_2–CH_2–CH_2–SO_3$
Methyl sufonate (S)	$–O–CH_2–CHOH–CH_2–CH_2–CHOH–CH_2–SO_3$

7.3.2.1 APPLICATION

The separation of protein molecules, peptides, and nucleotides based on the net charges they contain are most certainly the most widely employed chromatographic technology available. Proteins are composed of amino acids, each of which has a different total charge depending on the pH value of the fluid in which it is found. The amino groups carry a complete positive charge when the pH is acidic. The diagram shows that the carboxylic groups carry a complete negative charge when the pH is essential.

The pH value at which the magnitudes of the opposite charges on the protein are equal is known as the isoelectric point. Keeping a protein at pH values less acidic than the isoelectric point is accomplished using a cation exchanger.

The opposite is true of a protein that contains an anion exchanger, which will maintain a higher pH than the isoelectric point. Most of the proteins are negatively charged at physiological pH values (pH 6–8), anion exchange columns are almost exclusively utilized in protein purification. Amino acid separation occurs by either decreasing the net charge by lowering the pH or by increasing the ionic strength to break up the ionic interaction between the amino acids, depending on the situation. In addition, ion-exchange chromatography has been utilized in the large-scale separation of sugar, notably to enhance fructose while concurrently separating cane molasses, and has been employed to separate sugar at high temperatures.

Essentially, it is a technique that allows for the purification of a biomolecule to determine its biological function or particular chemical structure. To achieve separation and purification, this separation technique takes advantage of the unique property of particular natural interaction.

7.3.2.2 PRINCIPLE

A reversible complex that interacts with its analyte is formed and eluted by changing the buffer or adding a competitive ligand. Separation is based on the binding affinity of biomolecules with their respective biologically active ligands (this ligand is covalently bonded to the solid matrix). Fused proteins, monoclonal antibodies, and glycoproteins are all widely purified using affinity chromatography media, also known as affinity chromatography medium.

7.3.3 COMPONENTS OF AFFINITY CHROMATOGRAPHY

Beads, ligand, and a solution containing the substrate to be purified are among the primary components required for an affinity chromatography technique. A wash is also necessary to elute non-binding contaminants from the solution, as well as a final wash to elute the bound substrate from its ligand, both of which are detailed below. The matrix can be made of agarose, dextran, cellulose derivatives, and other polymers. Sepharose (agarose-based) is one of the most widely used because the hydroxyl groups on the sugar residues can be easily modified to accept a ligand; other polymers can also be used as the matrix; the ligand is chosen by the target molecule. For isolation of a specific enzyme from its environment, a ligand, a substrate analog, an inhibitor, or even a cofactor can be used. Antigen A can be used as a ligand in conjunction with the antiserum to separate antibodies specific for antigen A from an antiserum. For the technique to be successful, the substance to be separated must be capable of responding to a specific ligand added to an insoluble matrix. It's critical that the binding isn't too tight or too loose. The KD value (dissociation constant) should be between 10^{-4} and 10^{-8}.

Following the immobilization of the ligand-matrix gel in a column, a sample combination comprising the pure isolate is delivered to the immobilized ligand and allowed to react. Only proteins with a certain binding affinity for a ligand are kept in the binding matrix as proteins pass through a matrix of affinity beads. On the other hand, certain pollutants in the gel column remain unbound or relatively lightly bound. Several washes with varying pH, salt content, or temperature are passed through the gel to eliminate all pollutants, a solid wash must be used, but it should not be too strong that it removes the bound isolates from the solution. The protein mixture bound to the ligand-matrix in the gel is recovered using either specific or non-specific elution after contaminants have been removed. Non-specific elution can occur when the pH or ionic strength of the solution changes. The ligand's reversible binding properties are used in this second wash, allowing the bound protein to dissociate from the ligand. Before the isolation is completed, the pure protein is recovered in a buffered solution that may be contaminated with a specific eluting agent or contain high quantities of salts that can be removed using size exclusion chromatography (Figure 7.7).

7.3.3.1 APPLICATION

Purification of enzymes and other proteins, such as receptor proteins and immunoglobulins, can be accomplished using affinity chromatography. Affinity chromatography is used to separate and quantify glycosylated hemoglobins, fusion proteins, antibodies, and other proteins. Complementary RNA and DNA can be isolated using immobilized single-stranded DNA. Separation of amino-acyl-ribonucleic acid synthetase I on tRNA conjugate of sepharose matrix is accomplished using affinity chromatography.

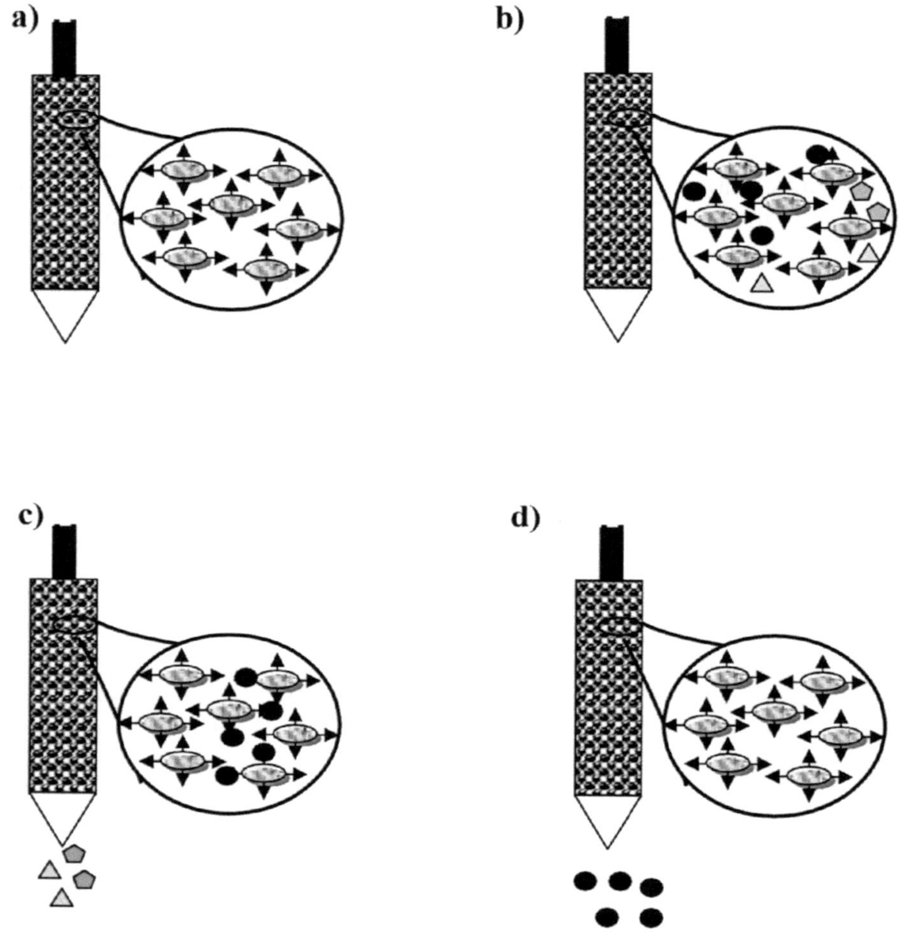

FIGURE 7.7 (a) Loading of ligand-matrix gel into elution column; (b) proteins with specific affinity bind to ligand; (c) unbound impurities get washed off; and (d) purified protein of interest.

7.3.4 SIZE EXCLUSION CHROMATOGRAPHY (GEL FILTRATION)

The method of size exclusion chromatography, commonly known as gel filtering, separates molecules depending on their size and shape. The molecular weight, shape, and size of the components in the sample mixture are used to separate them. The molecular weight of an unknown molecule can be determined using gel filtration. Gel filtration is an important step in the purification of proteins, polysaccharides, and nucleic acids.

7.3.4.1 PRINCIPLE

The stationary phase, which is in equilibrium with the mobile phase, is made up of porous glass granules or gel particles. Large molecules are entirely blocked from entering the pores. They elute first because they have only access to the mobile phase between both beads, whereas molecules of typical size are only substantially included in the elution. Some of the openings in the beads can be slipped into, but not all of them are available. These molecules elute in the elution gradient between the large (excluded) and small (fully included) molecules. Small molecules, on the other hand, are completely incorporated. They elute last in a gel filtration separation after being dispersed between the mobile phase inside and outside the molecular sieve (Figure 7.8).

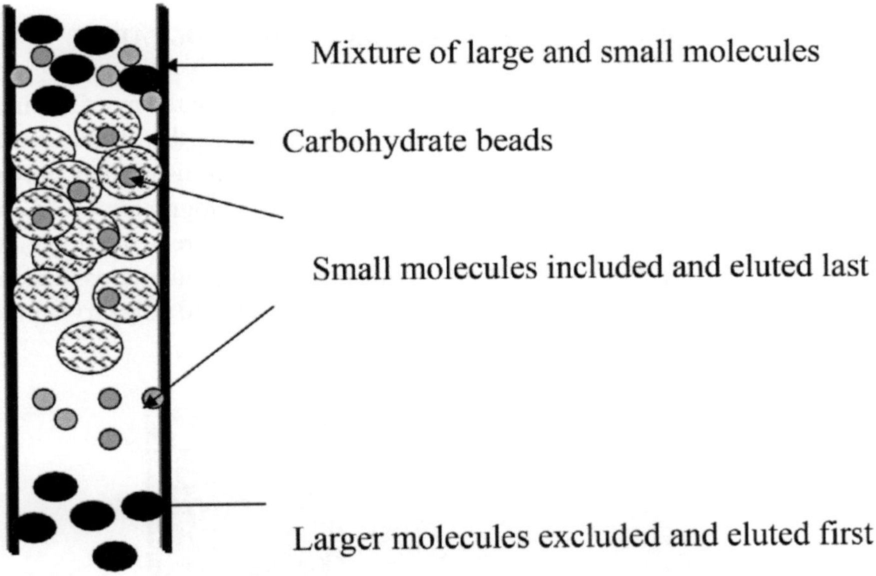

FIGURE 7.8 Size exclusion chromatography.

In gel filtration, the separations are described by the following equation:

$$V_r = V_0 + K_d V_i$$

where; V_r is the retention volume of the protein; V_0 (void volume) is the volume of mobile phase between the beads of the stationary phase inside the column; V_i (included volume) is the volume of mobile phase inside the porous beads; and K is the partition coefficient; the extent to which the analyte can penetrate the pores in the stationary phase, with values ranging between 0 and 1. If the analyte is larger and completely excluded from the mobile phase within the gel, $K_d = 0$, whereas if the analyte is sufficiently small to gain complete access to the inner mobile phase, $K_d = 1$. The variation of K_d between these two limits makes possible the separation of analytes within a narrow molecular size range on a given gel.

For two substances of different relative molecular masses and K_d values, K'_d and K''_d and the difference in their elution volumes, V_s, can be given as:

$$V_s = (K'_d - K''_d)\, V_i$$

where V_i is the inner volume within the gel available to a compound whose $K_d = 1$.

7.3.5 COMPONENTS OF SIZE EXCLUSION CHROMATOGRAPHY

The basic components of the size exclusion experiment are the matrix, chromatography column and the elution buffer. The matrix is the porous bead, which is the separation medium, for example, agarose porous bead as shown in Figure 7.9. It is the stationary phase of the gel chromatography. There are different types of gel filtration matrices available in different fractionation ranges (Table 7.4). Matrices that are commonly used include cross-linked dextrans (e.g., Sephadex), agarose (Sepharose, Bio-Gel) and polyacrylamide (Sephacryl, Bio-Gel P).

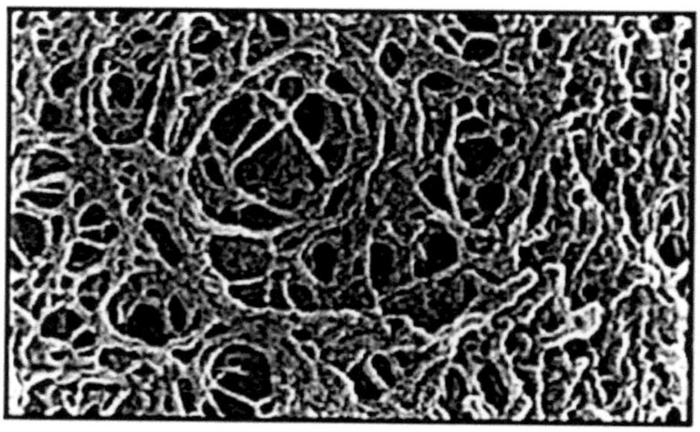

FIGURE 7.9 Porous bead structure of size exclusion chromatography. (Scanning electron micrograph taken from the Department of Biochemistry, Uppsala University, Uppsala, Sweden).

TABLE 7.4 Matrices Commonly Used for Size Exclusion Chromatography

Matrix	Trade Name	Fractionation Range for Globular Protein (Da)
Dextran	Sephadex	$7 \times 10^2 - 3 \times 10^4$
Agarose	Sepharose	$1 \times 10^4 - 4 \times 10^4$
–	Bio-gel A	$1 \times 10^4 - 15 \times 10^6$
Polyacrylamide	Bio-Gel P	$1 \times 10^2 - 1 \times 10^5$

The fractionation range of a matrix refers to the range of molecular weights from which it can separate them. Consider the case of a matrix with a fractionation range (expressed in molecular weight) ranging from 1 to 100 kilo Daltons (Kda), for example. In the range of 1 to 100 Kda, molecules will penetrate the beads with variable degrees of efficiency and be segregated from one another fully or in part by the beads. Molecules with a molecular weight greater than 100,000 nm will be eluted into the reactor's void volume rather than entering the beads. Because each molecule penetrates the beads completely and equally, molecules with an average molecular weight of 1,000 or less are indistinguishable from the others. For elution, these compounds require the maximum amount of buffer available.

To enhance the amount of stationary phase and thus pore volume, exclusion chromatography columns are often longer than those used in other types of chromatography. The dispersed phase of chromatography is the elution buffer, which is responsible for transporting the material through the matrix and out the column. The elution buffer develops the column using the matrix and applied sample. The buffer flow transports the molecules from the sample into the matrix, where they are gradually separated. A single mobile phase and isocratic elution are required for exclusion chromatography. The segregated zones of molecules that flow out of the column are collected for analysis, with UV absorption spectrophotometric detectors being the most frequent method.

7.3.5.1 APPLICATION

Size exclusion chromatography is often used to purify biomolecules based on their size, such as proteins, peptides, hormones, antibodies, and polysaccharides. This technique can be used for more than just purification; it can also be used to measure relative molecular mass (Mr). Using a Sephadex G-25 column, high molecular mass solutions can be desalted. The protein-ligand binding has also been studied using exclusion chromatography.

7.4 ULTRAFILTRATION

Ultrafiltration is a type of filtration that uses a pressure gradient to drive a membrane separation process. The dissolved components of a liquid are fractionated by the membrane based on their solvated size and structure. Ultrafiltration is most typically employed to process a solution containing a variety of molecularly different components. Ultrafiltration membranes allow solutes to pass based on the membrane's nominal molecular weight cut off (NMWCO).

7.4.1 PRINCIPLE OF ULTRAFILTRATION

Ultrafiltration is accomplished through the use of a porous semi-permeable membrane. The solvent and small molecules are forced through the membrane to the permeate under pressure (driving force), and macromolecules larger than the pore size are trapped by the membrane and concentrated in the retentate reservoir. Permeation is the term used for the membrane mass transport mechanism, which is defined by Fick's Law in simple words:

$$J = \mathrm{P}(\Delta \mathrm{DF}/l)$$

where J = flux; P = permeability coefficient; l = membrane thickness; ΔDF = driving force across the membrane.

7.4.2 FLOW MODE IN FILTRATION

Membrane filtration can be done in a cross flow or dead-end flow configuration. The feed flow in dead-end filtration (traditional filtration) is perpendicular to the membrane surface (Figure 7.10(a)), causing debris build-up at the membrane surface, which results in membrane blinding and, as a result, reduced fluid permeation. Cross flow filtration (tangential flow) on the other hand, produces a continuous turbulent flow (Figure 7.10(b)) and reduces the formation of a membrane fouling layer due to feed stream debris and macromolecules. Particles deposited on the membrane surface are whisked away by the feed flow, preventing the membrane from being clogged. Cross flow mode, as compared to dead-end filtering, increases steady, long-term productivity. Many factors influence crossflow filtration, including crossflow velocity, transmembrane pressure, membrane resistance, cake layer resistance, suspended particle size distribution, and particle shape.

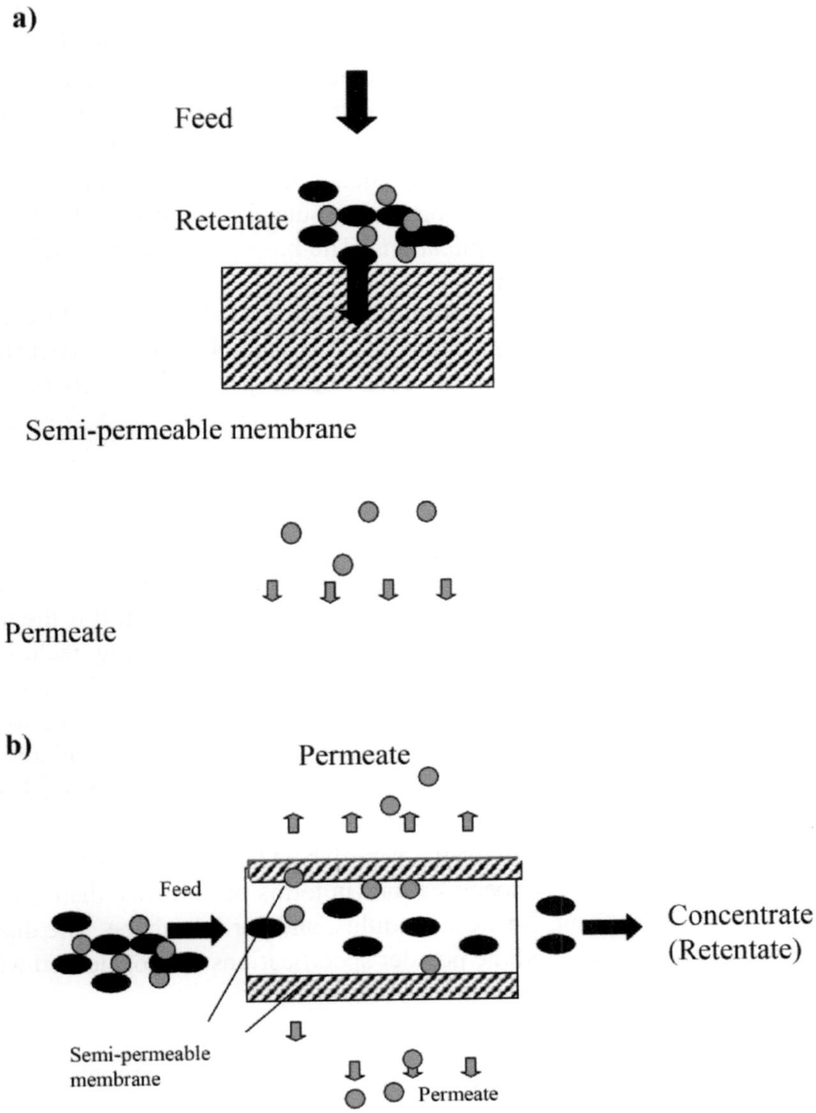

FIGURE 7.10 Schematic diagram showing – (a) dead-end; and (b) crossflow mode in ultrafiltration.

7.4.2.1 APPLICATION

In several engineering and applied scientific domains, ultrafiltration is required to extract colloids and particulate materials from liquid solutions. This separation process is used for a wide range of applications, including biological macromolecule processing, removal of suspended particles and pathogens from water and wastewater treatment, food processing, beverage and cosmetic industry, bacterial cell harvesting, plasma separation from whole blood, and dewatering of suspended solids such as mineral slurries. Food and biotechnology applications contribute for about 40% to 44% of all ultrafiltration membrane applications. The bioseparation of proteins is a crucial part of this application.

The ultrafiltration method is primarily used in the food sector to extract key flavoring compounds from materials that cannot be heat pasteurized. Ultrafiltration is used in modern biotechnological processes to separate cell mass, proteins with varying molecular weights, and so on.

7.5 SPRAY DRYING

Spray drying is a multiple process that involves spraying droplets, drying them, and then removing the dried powder from the air stream. It is the most extensively utilized industrial particle production and drying technique. Spray drying has numerous uses in the pharmaceutical and food industries. The pharmaceutical spray dryer is a versatile, modern, and simple-to-use drier with a variety of control mechanisms. It powders, granulates, or agglomerates solutions, suspensions, and emulsions. It can be used as a laboratory drier for research and development or as part of a batch production process. As a result, spray drying is preferred when the finished product must meet stringent quality specifications in terms of pore size distribution, residual moisture content, fineness modulus, and particle shape. The drying properties of the product and the powder specifications are considered while selecting the operating settings and dryer design.

7.5.1 SPRAY DRYING SYSTEM

As shown in Figure 7.11, the spray drying system consists of a nozzle atomizer, feed pump, air heater, air disperser, drying chamber, and exhaust air cleaning and powder recovery methods.

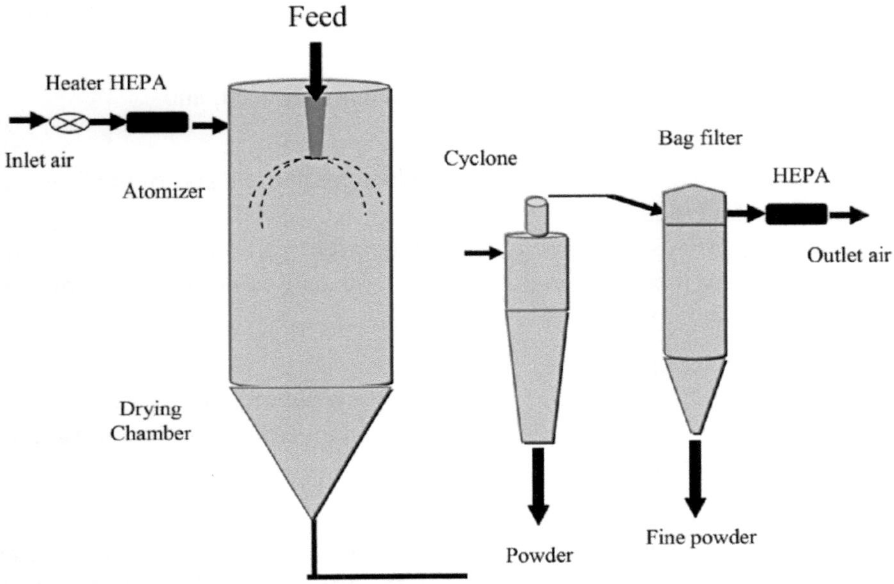

FIGURE 7.11 Schematic diagram of typical spray drying system.

7.5.2 OPERATING PRINCIPLE

Every spray drying process consists of the following unit operations:

1. **Pre-Concentration of Liquid:** To lower the water content of the liquid feed, it is usually pre-concentrated through evaporation.
2. **Atomization:** In this process, droplets are formed from a liquid feedstock. Spray drying is pumping a liquid product concentration through an atomizing device, where it is broken down into tiny droplets, which then come into contact with hot air in a drying chamber. Atomizers that spray droplets are either rotary or nozzle atomizers.

Spray dryers are equipped with atomizing machines, distinguishing spray drying from other drying methods. They give an extremely huge surface area for exposure to drying pressures, allowing for faster drying times. This is a significant metric because it responds quickly to changes in the system and accurately represents the quality of the end product. As a result, it is vital to monitor the temperature of the outflow air. It should be adjusted to a sufficiently high temperature to produce the desired moisture while avoiding heat damage. Adjusting the temperature of the outflow air can be accomplished through the use of two controls:

- Modifying the rate at which the feed is delivered; and
- Adjusting the inlet temperature.

Unless heat damage occurs before drying, the particle size of the finished product must be reduced.

Smaller particles dry more quickly than larger particles, and as a result, they suffer only modest thermal damage. This can be performed primarily through three methods:

- A smaller orifice size;
- Increase the atomizing pressure of the solution; and
- Improve viscosity by increasing feed temperature or lowering solids content in the feed.

3. **Airflow:** When spray droplets generated during the atomization process come into contact with a stream of hot air, they quickly lose moisture while still suspended in the drying air. The point of contact between spray droplets and drying air controls evaporation processes and product temperatures in the dryer. There are three main modes of communication:

i. **Co-Current Spray Drying:** A variable speed progressive cavity pump pumps the spray droplets to the top of the drying chamber, ensuring a highly smooth flow and steady flow rate. The drying air and particles always travel in the same direction through the drying chamber (Figure 7.12). Hot air is pumped into the room and onto the droplets, causing the surface moisture to evaporate quickly. At the time the particles fall to the discharge point at the base of the cone, the moisture in the particles (bound form of moisture) is evaporated. In this configuration, the particle temperature never exceeds the exhaust air temperature, which is an ideal model for drying heat-sensitive products. The additional advantage of co-current spray drying is that it can handle both low and high viscosity inputs simultaneously. When using the co-current two-fluid nozzle, powders are generated with an average particle size in the range of 10–40 microns.

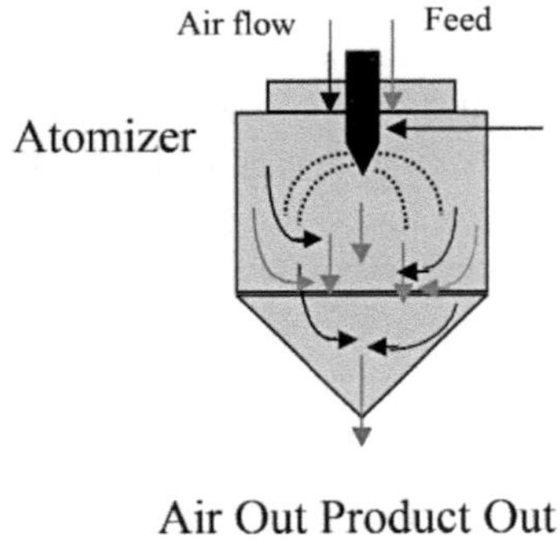

FIGURE 7.12 Co-current spray flow dryer.

ii. **Counter-Current Spray Drying:** Figure 7.13 depicts the movement of drying air and particles through the drying chamber in the opposing directions of each other. If you are using a counter-current design, the nozzle should be placed at the base of the chamber and the hot air should be sprayed upwardly into it. This mode is appropriate for materials that require some degree of heat treatment throughout the drying process. The powder temperature exiting the dryer is often higher than the temperature of the exhaust air exiting the dryer.

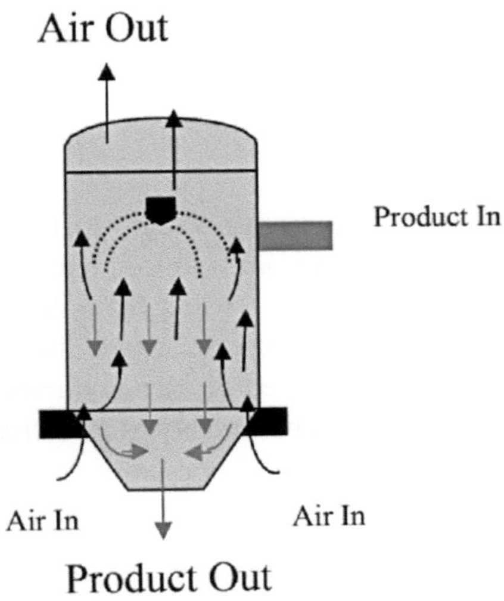

FIGURE 7.13 Counter current flow dryer.

iii. **Mixed Flow (Fountain Mode):** The passage of particles through the drying chamber is characterized by both co-current and counter-current phases (Figure 7.14). In this mode, heat-stable products with coarse powder requirements can be dried using nozzle atomizers that spray upwards into an incoming airflow. Heat-sensitive products can be dried using atomizers that spray downwards into an integrated fluid bed. The air inlet and outlet are located in the top portion of the drying chamber, makes it particularly suitable for drying heat-stable products. The fountain mode benefits from spray drying heavy solids feeds and creating free-flowing powders with mean particle sizes in the range of 60–100 m, which is ideal for various applications.

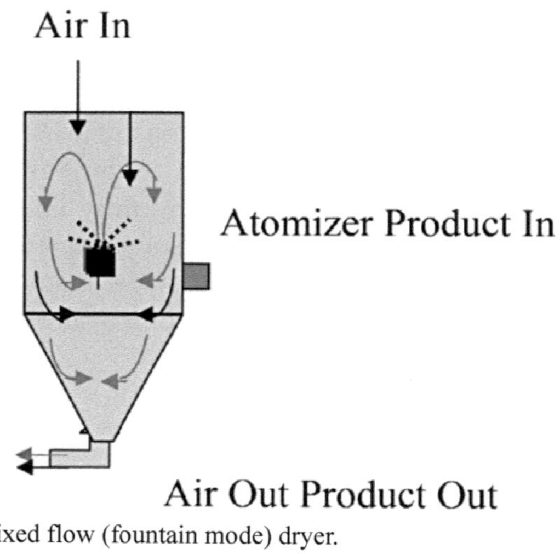

FIGURE 7.14 Mixed flow (fountain mode) dryer.

4. **Powder Recovery:** Maximum powder must be recovered from the air stream for both economic and environmental reasons. The collection system separates the dry powder from the moist air. Bag filters are very efficient, but labor expenses, sanitation, and the risk of heat damage owing to the extended residence durations make them less acceptable. Although cyclones are not as effective as bag filters, they can be used in series. Drying huge amount of powder per hour is possible with cyclone dryers. Centrifugal motion separates the dry powder from the moist air in cyclones. When a mixture of particles and air enters the cyclone system, it causes a large rise in airspeed, which causes centrifugal action. The heavier and denser powder

particles are squeezed into the cyclone walls, while the lighter moist air is propelled out the exhaust pipes. The powder settles at the bottom of the cyclone and is discharged via a discharging gear. The most cost-effective outlet air cleaner is a wet scrubber. By spraying the wash stream through the air, it dissolves any dusting powder left in the air stream into either water or the feed stream.

5. **Cooling System:** The air-conveying ducts for the dry powder are coupled with cooling systems in some spray drying systems, allowing cold air to carry the product through conveying pipes.

Open cycle and close cycle spray dryers, as depicted in Figures 7.15(a) and 7.15(b), are the two most popular spray dryers. Using cyclones and scrubbers, the open process warms and cleans the incoming air. After drying the food product, the hot air is squandered and returned to the environment. Since the heated air is recycled, the closed-loop system is more efficient.

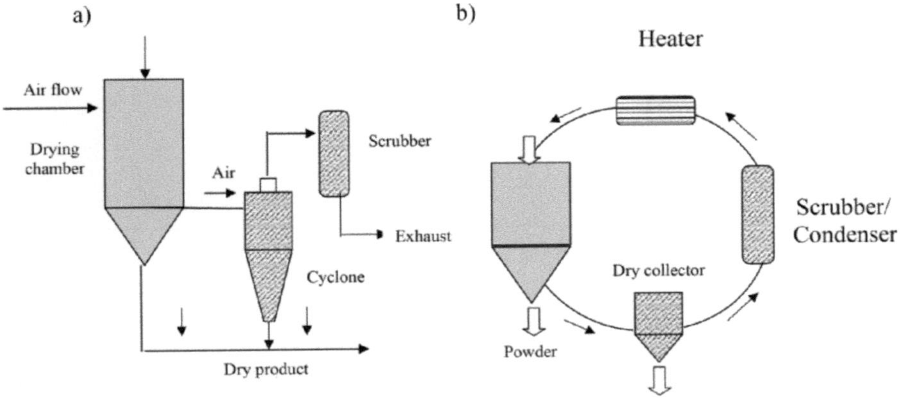

FIGURE 7.15 (a) Open cycle spray drying layout; and (b) closed cycle spray drying layout.

7.5.3 APPLICATION

In the pharmaceutical and food industries, the spray drying method offers a wide range of uses. Spray drying is the most popular process for creating microencapsulated formulations of active medicinal components such as peptides, proteins, and DNA in the pharmaceutical industry. Spray-drying can be used to make biodegradable microparticles such as dl-lactic/glycolic acid microparticles for long-term release of a water-soluble drug (Thyrotropin-releasing hormone: TRH). In the food industry, it's critical to dry heat-sensitive products like fruits and vegetables. Spray drying, for example, appears to be the most efficient method for creating high-quality tomato powder at a cheap cost. Another fascinating application of spray drying is the manufacturing of skim milk powder. Throughout the pasteurization process, a low pasteurization temperature and careful treatment are necessary. Two-stage drying is recommended for this purpose. Because the particle temperature is substantially lower, especially during the critical drying period from 20 to 10% moisture, it allows for a more careful drying, resulting in a powder without protein denaturation.

KEYWORDS

- **chromatography**
- **freeze-thawing**
- **gel filtration**
- **ion exchange chromatography**
- **osmotic shock**
- **spray drying**
- **ultrafiltration**

REFERENCES

Belter, P. A., Cussler, E. L., & Hu, W. S., (1988). *Bioseparations: Downstream Processing for Biotechnology.* John Wiley & Sons, New York.

Brown, P. R., Grushka, E., & Lunte, S., (2005). *Advances in Chromatography* (Vol. 43). Marcel Dekker, Inc., New York.

Doran, P. M., (2004). *Bioprocess Engineering Principles*. Elsevier Academic Press, London.

Drioli, E., & Giorno, L., (1999). *Biocatalytic Membrane Reactors: Applications in Biotechnology and the Pharmaceutical Industry*. Taylor & Francis Ltd., London.

Flickinger, M. E., & Drew, S. W., (1999). *Encyclopedia of Bioprocess Technology: Fermentation, Biocatalysis and Bioseparation* (Vol. 1–5). John Wiley & Sons, New York.

Ghosh, R., (2003). *Protein Bioseparation Using Ultrafiltration: Theory, Application and New Developments.* Imperical College Press, London.

Hage, D. S., (2005). *Handbook of Affinity Chromatography* (2nd edn.). CRC Press.

Joshi, V. K., & Pandey, A., (1999). *Biotechnology: Food Fermentation (Microbiology, Biochemistry and Technology)* (Vol. 2). Educational Publishers & Distributors, New Delhi.

Mathews, C. K., (2000). *Biochemistry* (3rd edn.). Addison Wesley Longman, Inc., San Francisco, CA.

Pisecky, J., (2004). *Spray Drying in the Cheese Industry*. IDF Symposium on Heese. Prague, Czech Republic.

Rousseau, R. W., (1987). *Handbook of Separation Process Technology*. John Wiley & Sons, New York.

Stanbury, P. F., Whitaker, A., & Hall, S. J., (1997). *Principle of Fermentation Technology*. Butterworth- Heinemann Ltd., Oxford.

Swarbrick, J., & Boylan, J. C., (2001). *Encyclopedia of Pharmaceutical Technology* (Vol 20). Marcel Dekker, Inc., New York.

Vijayalakshmi, M. A., (2002). *Biochromatography: Theory and Practice*. Taylor & Francis, London.

Wang, W. K., (2001). *Membrane Separation in Biochemistry* (2nd edn.). Marcel Dekker, Inc., New York.

Weatherley, L. R., (1994). *Engineering Processes for Bioseparations*. Butterworth-Heinemann Ltd., Oxford.

Wilson, K., & Walker, J., (2000). *Practical Biochemistry: Principle and Techniques* (5th edn.). Cambridge University Press, UK.

CHAPTER 8

Principles of Overproduction of Metabolites

8.1 INTRODUCTION

Metabolites are the substances that cells use to maintain life. Metabolite is a group of small molecules of chemical transformations which include all of the compounds used in metabolism. The numerous metabolites have different roles like fuel, structure, stimulus, regulators/inhibitors in enzymes, catalyst in their self-defense system, and association with other living beings. These metabolites, associated with the growth, development, and reproduction of organisms, are primary metabolites, like ethanol, lactic acid (LA), and certain amino acids. In contrast, secondary metabolites are generally organic compounds that play a significant role in defense systems and are not directly involved in those processes mentioned for primary metabolites. Secondary metabolites are essential for humans and are used as medicines, flavorings, and recreational drugs. Secondary metabolites produced by microorganisms have a significant impact on human health, nutrition, and economics. These metabolites encompass a wide range of compounds, including immune-modulating agents, antibiotics, enzyme inhibitors, receptor antagonists and agonists, pigments, antitumor agents, pheromones, and growth regulators for animals and plants. Their diverse effects contribute to their importance in various fields. Three categories of microbial world, i.e., Eukarya, Archaea, and Bacteria are known for having almost all ecological niches of the earth. These microorganisms are characterized by their ability to generate these secondary metabolites. Although secondary metabolites are not directly connected with their growth and reproduction, they are vital in stress management in different conditions. Eventually, most of these secondary metabolites have a wide range of bioactivities and are explored in various

Handbook of Industrial Food Microbiology. Manju Nehra & Vikash Nain
© 2025 Apple Academic Press, Inc. Co-published with CRC Press (Taylor & Francis)

food, agro-chemical, and pharmaceutical industries. There are some common and basic pathways associated with the production of these metabolites. Based on these pathways; they are categorized in various groups, such as non-ribosomal peptides, terpenoids, alkaloids, and polyketides. They have found a remarkable place in market as agricultural chemicals, chemo-therapeutic, and antibiotics compounds without exerting any significant effect on environment with high efficiency and low toxicity. These searches have gained immense attention due to the increased resistance of infectious microorganisms, e.g., Streptococcus, Mycobacterium, Staphylococcus on pre-existing drugs.

In the recent decade, applications of biotechnological tools have opened broad scope of these metabolites in their production from efficient, potential, and beneficial micro-biomass. Extremophilic and plant-associated microbes have developed essential secondary metabolites for various applications.

8.2 IMPORTANCE OF SECONDARY METABOLITES

Secondary metabolites, including antibiotics, serve a variety of functions:

- Competitive tools against other living organisms including pests, and microbes, plants, and animals;
- Metal carrier;
- Symbiotic agents (organisms that live in a symbiotic relationship with others);
- Reproductive agent;
- Differentiation effectors;
- Agents of communication between organisms.

Production of secondary metabolites from microorganisms:

1. **Liquid Fermentation:** Secondary metabolites are obtained through batch or fed-batch culture in submerged fermentation. Inoculum is prepared after careful strain improvement of the producing microbes. In this process, shake flasks culture technique is used at initial stage, then culture having active growth phase is shifted to small fermenter. Later, it is redirected to a large fermenter having the production media. Several parameters are regulated for efficient production, such as medium composition, pH, temperature, agitation, and aeration level. Suitable inducers are employed to produce different metabolites such as methionine is introduced for cephalosporin fermentations,

phosphate is constrained in chlortetracycline fermentations, and glucose is avoided in penicillin or erythromycin fermentations.

2. **Solid-State Fermentation (SSF):** It is characterized as a microbial culture that occurs on the surface and inside a solid matrix in the absence of free water, and has significant potential for secondary metabolite production. Two forms of solid-state fermentation can be categorized based on the solid phase used:

   i. Solid culture of one support-substrate phase solid phase; and
   ii. Solid culture of two substrate-support phase solid phases.

Solid-state fermentation has several benefits over submerged fermentation, including lower energy requirements as oxygen is transmitted directly to the microorganism. Non-requirement of sterile conditions, high yield, and short fermentation period are other advantages.

8.3 OVERPRODUCTION OF SECONDARY METABOLITES

There are some methods/techniques which have an inhibitory or promoting effect on the production rate of the secondary metabolites. Some of them are summarized here.

1. **By Metabolic Engineering:**

   i. **Induction:** Some substances in secondary metabolite fermentation have stimulatory effects similar to enzyme induction. The function of tryptophan in *Claviceps* sp. ergot alkaloid fermentation is a good example. Although the amino acid is a precursor, its function as an inducer of some enzymes required for alkaloid biosynthesis appears to be more critical. This is because tryptophan analogs, in addition to not being absorbed into the alkaloid, often stimulate the enzymes involved in the alkaloid's biosynthesis. Additionally, tryptophan must be incorporated during the growth phase to maximize the alkaloid synthesis; otherwise, it got reduced significantly. It will suggest that the trophophase contains certain biosynthetic enzymes or chemical reactions that lead to alkaloid transformation, establishing a relation between the idiophase and the trophophase. Methionine seems to exert an identical induction in the cephalosporin C production from *Cephalosporium acremonium*.

ii. Catabolic Regulation:

a. **Carbon Source Regulation:** Carbon has long been believed to play a role in secondary metabolism regulation. For a long time, it has been believed that penicillin is not released in a glucose-containing medium until the glucose is exhausted, at which point the idiophase starts. The same result has been found in the synthesis of cephalosporins. In fact, "glucose effect" is associated with large number of secondary metabolites, presence of sugar has a limiting effect on the production of secondary metabolites. Exhausting of easily available carbon source before utilizing the less available source is known as glucose effect. But it will be a misleading term as, in the absence of glucose, other carbon may be preferred where multiple carbon sources are available. *Mortierella* sp. produces the most β-carotene on fructose, even though galactose is a better carbon source for growth. Carbon sources that have been found suitable for secondary metabolite production include sucrose (tetracycline and erythromycin), soybean oil (kasugamycin), glycerol (butirosin), starch, and dextrin (fortimicin). It's relatively simple to determine if the catabolite inhibits or represses synthesis. The synthesis of the enzymes required for the metabolite's synthesis is inhibited in catabolite repression. It is proven corrected that production got severely reduced by adding the test substrate just before the beginning of secondary metabolite synthesis. Hence, regulations of test substrates play a significant role in the synthesis of secondary metabolites.

b. **Nitrogen Catabolic Regulation:** It has been observed to play an important role in both primary and secondary metabolism. It suppresses the enzyme synthesis which acts as glucose effect in carbon source. Here, nitrogen-containing substances (proteases, ureases, etc.) are used when the easily utilizable nitrogen sources (ammonia) are exhausted. Soybean act as a preferred source of nitrogen in the production of streptomycin by fermentation like lactose works in penicillin production, i.e., slow utilization of the component. Tri-hydroxy toluene production by *Aspergillus fumigatus*, bikaverin production by *Gibberella fujikuroi*, and cephamycin production by *Streptomyces* spp. are some secondary metabolites which are influenced by nitrogen catabolic regulation. In these examples, nitrogen must be depleted before the initiation of secondary metabolite generation.

iii. **Feedback Regulation:** Feedback inhibition is a common regulatory mechanism that governs primary metabolite biosynthesis. The first biosynthetic stage is normally inhibited by an elevated concentration of the last metabolite in the metabolic pathway. In several studies, feedback regulation occurs in secondary metabolism, in which the product prevents its further synthesis. Feedback control seems to prevent microorganisms that produce antibiotics from overproduction through inhibiting essential enzymes involved in their biosynthesis (Demain, 1974). Auxotrophic mutants were used to maximize the synthesis of secondary metabolites such as tetracyclines, streptomycins, kojic acid, penicillins, actinomycins, etc. (Demain, 1973). Lein (1986) identified biosynthetic control mechanisms and their disintegration in studies to improve penicillin-producing strains.

Penicillin inhibition by lysine is an example of feedback control. Since penicillin and lysine are end-products of a branched pathway, feedback inhibition by L-lysine influences penicillin biosynthesis in *Penicillium chrysogenum*. The primary enzyme in the process, homocitrate synthetase, is inhibited by lysine feedback, which prevents the formation of α-aminoadipate. The inhibitory effect of lysine is eliminated when α-aminoadipate is added. The biosynthesis of cycloheximide and nystatin in *Strefitomyces noiirsei* is another example of feedback inhibition regulating secondary metabolite biosynthesis. Aromatization of the dimethyl cyclohexanone ring inhibited the overproduction of toxic cycloheximide and developed biologically inert actiphenol.

Elander & Vournakis (1986) represented an antimetabolite's positive effect on producing a secondary metabolite. In *Pseiidoitionas fliioresceiis*, tryptophan is required as a precursor in the biosynthesis of pyrrolnitrin. By selecting a mutant strain resistant to tryptophan analogs, it was possible to obtain a strain that did not require tryptophan in the cultivation medium while still producing high concentration of the desired antibiotic pyrrolnitrin. In general, any microorganism that produces toxic compounds has a variety of mechanisms that inhibit its "suicide" due to over production of such compounds. Many secondary metabolites or analogs have identified a feedback mechanism that inhibits their production. Antibiotics such as Odorox, which is used in poultry feeds and is effective against Gram-positive bacteria, chloramphenicol, penicillin, cycloheximides, and 6-methyl salicylic acid are examples of self-feedback inhibition mechanism.

iv. **Energy Charge Regulation of Secondary Metabolites:** Secondary metabolism has a significantly lower tolerance for inorganic phosphate concentrations than primary metabolism. Prokaryotic and eukaryotic organisms can grow well in a range of inorganic phosphate concentrations of 0.3–30 mM while the same ranges between 1 and 10 mM for the production of secondary metabolites. Antibiotic production is inhibited by high phosphate levels. As a result, the antibiotic industry chooses low-phosphate media or decreases phosphate content by incorporating phosphate-complexing agents to the medium. Phosphate promotes fast respiration rates, DNA and RNA synthesis, and glucose consumption, forcing the growth process to switch from idiophase to trophophase. This shift was observed to happen irrespective of the growth stage of the organisms. As a result, phosphate exhaustion assists in the onset of idiophase. Another theory is that a high phosphate level causes carbohydrate catabolism pathways to transfer from HMP to EMP, preferring glycolysis.

v. **Stimulatory Effect of Precursors:** The inclusion of precursors boosts productivity and increases yields in many secondary metabolite fermentations. In the early days of penicillin fermentation, adding phenylacetic acid found in steep corn liquor increased penicillin production. Methionine is needed for the experimental synthesis of aflatoxin by *Aspergillus parasiticus*. L-citrulline is an intermediate in the development of mitomycin by *Streptomyces verticillatus*.

vi. **Inorganic Compounds:** Phosphate and manganese are two inorganic compounds that profoundly impact secondary metabolite fermentation. The impact of inorganic phosphate has already been addressed. In brief, high phosphate levels stimulate growth, but it is detrimental to the development of secondary metabolites. Manganese, on the other hand, promotes the development of idiophase products especially in *bacilli*, such as bacillin, bacitracin, mycobacillin, subtilin, D-glutamine, protective antigens, and endospores. Interestingly, the amount required to promote growth varies from 20 or more times than the amount required to promote secondary metabolite production.

vii. **Temperature:** Generally, temperature required for the healthy growth (in trophophase) of micro-organisms varies in the range of 25–35°C while for the production of secondary metabolites it is very low compared to ideal growth temperature. Usually, secondary metabolites are favorably produced at 5–10°C. Hence, maintenances of proper temperature at different stages of fermentation significantly impact the production rate of secondary metabolites. Multi-stage temperature control is required for the efficient growth and production of secondary metabolites. The idiomorphic process happens to be slightly at higher temperature, whereas the trophic phase, which takes place later, requires a little lower temperature.

2. **By Genetic Engineering Techniques:** Genetic engineering-based overproduction of various secondary metabolites is governed by those genes which directly participate in the biosynthesis of these metabolites also known as structural gene along with regulatory and other genes involved in the production of primary metabolites and some antibiotic resistance genes. Increased yields of the target metabolite reduced undesirable co-metabolites, increased use of low-cost carbon and nitrogen supplies, and modification of cellular structure to a shape best suitable for separation of the mycelium from the substance and/or improved oxygen transport in the fermenter are all benefits of improved strains.

Genetic engineering-based techniques involved in secondary metabolite overproduction are further categorized into two groups, i.e., Classical genetic methods and Molecular genetic improvement methods.

8.4 CLASSICAL GENETIC METHODS

There are three methods for genetic methods: (1) mutation and random selection; (2) mutation and rational selection; and (3) genetic recombination methods.

8.4.1 MUTATION

Increasing the production of primary and secondary metabolites through the development of mutant strains is most common practice followed in the different organisms. Mutation sometime induces the positive effect and develops a much better strain in the production of secondary metabolites. Production of some antibiotics has increased severalfold (15–400 times) though the mutation, compared to their wild strains. The mutation may be of several types, such as:

i. **Mutation and Random Selection:** Using spontaneous mutations and random screening, new mutants are selected based on the controlled fermentation testing results. Various mutants such as UV-light, N-methyl-N'–nitro-N nitrosoguanidine and ethyl ethane sulfonate carry spontaneous and induced mutation in different strains. This approach has several advantages. It is easy to perform and flexible. There is no requirement for advanced instruments in this technique. This classical mutation method benefits fast productivity and low mechanical manipulation. Time-consuming processes and laborious nature are a few limitations of this approach.

ii. **Mutation and Rational Selection (Directed Selection Techniques):** The steps of this technique are:

   - Selection for a single attribute of the appropriate genotype differs from the one of ultimate interest but is easy to identify.
   - All unnecessary genotypes are discarded, allowing efficiently analyzing a large number of isolates.

   Requirements: A fundamental understanding of product metabolism and pathway regulation is expected.

   For example, when a toxic precursor of penicillin is added to the agar medium of penicillin-producing microorganisms, sensitive strains cannot develop, and only tolerant mutants with higher penicillin production can survive. In another study, the addition of rose Bengal and thymol to the medium of carotenoid-producing yeast subjected to visible light was used to select high carotenoid-producing strains.

iii. **Recombination Methods:** Protoplast fusion of closely related species of micro-flora having the desirable genetic material that have high production of secondary metabolites with new isolate with a lower level of production of new secondary metabolites can lead to the higher productivity of newly discovered secondary metabolites from both such strains. This technique is known as the recombination method. Here, genetic materials of two closely related strains with different production rates are fused to obtain higher productivity.

8.5 MOLECULAR GENETIC IMPROVEMENT METHODS

It is essential to have sufficient expertise and be capable of performing molecular genetic improvement techniques, including genetic recognition of the biosynthetic pathway, well-studied protocols and adequate vectors. Some important techniques used for the enhancement of secondary metabolites production rate in different strains are discussed here:

1. **Amplification of Gene Responsible for the Production of Secondary Metabolites:** This scheme is alienated into two methodologies:

   i. **Targeted Gene Duplication (or Amplification):** There must be a chromosome segment where genetically modified genes may be added that doesn't affect the fermentation. In the second step, this neutral site is then cloned multiple times and inserted in chromosomal DNA by homologous recombination technique. Upon this transformation, the targeted gene is inserted into the chromosomal neutral site using the same technique. Successful transfer of targeted genes may enhance the production rate of secondary metabolites in that organism. *tyl*F gene is responsible for O-methylation production in tylosin biosynthesis from S*treptomyces fradiae.* Targeted gene duplication of the mentioned gene enhanced the 60% production of tylosin compared to normal strain. This is achieved by inserting the two copies of *tyl*F gene.

ii. **Whole Pathway Amplification:** Here, the complete biosynthetic pathway of a particular secondary metabolite is recognized and amplified with the same technique. Complete pathway analysis is required in this method. So, this technique is limited to only those metabolites whose biosynthesis pathways are well studied earlier.

2. **Inactivation of Competing Pathways:** This technique involves blocking a pathway that competes for a typical intermediate key precursor such as cofactors and reduces power and energy supply. This strategy can improve strains' ability to channel precursors to SM biosynthesis. These strategies include transposon mutagenesis, chromosome destruction and the insertion of an antisense synthetic gene in actinomycetes. α-aminoadipic acid, for example, is a substrate to penicillin biosynthesis that often serves as a branching point for lysine synthesis. The lys2 gene, which binds with α-aminoadipic acid to lysine, has been disrupted, resulting in auxotrophic amino acid with a 100% rise in penicillin production rate.

3. **Amplification or Disruption of Regulatory Genes:** Molecular regulation is more difficult than defining the biosynthetic mechanism and cloning the genes associated. For example, amplifying a regulatory gene (ccaR1) resulted in a threefold increase in the synthesis of β-lactam compounds in *Streptomyces clavuligerus*. When negatively acting regulatory gene mmy of methylenoomycin biosynthesis was disrupted, the production of Actinorhodine increased seventeen-folds. In another study, a single copy of the positively acting gene act II was introduced into *Streptomyces coelicolor*, resulting in a 35-fold increase in actinorhodin development.

4. **Combinatorial Biosynthesis:** Novel antibiotics are being developed by utilizing nonconventional substances as a substrate for the micro-organism's biosynthetic enzymes. These enzymes can be mutated or manipulated to have a higher affinity for certain non-natural substrates. Using genetic engineering, different action modules of enzymes such polyketide synthases may be rearranged to construct a microbial strain that synthesizes an antibiotic with novel properties. For example, expression of *A. orientalis* gene responsible for glycosyltransferase transferred in *Streptomyces toyocaensis* (producer of the nonglycosylated hepta-peptide) results in a novel mono-glycosylated derivative.

8.6 THROUGH THE APPLICATION OF SYNTHETIC BIOLOGY

Synthetic biology can produce unnatural substances with standardized characteristics by applying the engineering theory of systems design to biological systems. A diverse library of biological modules is needed for the effective implementation of this technology. These modules may be combined to create complex mechanisms that can be programed in the desired way. For the efficient production of new compounds, the assembled pathway requires a suitable surrogate host, the most typically *Escherichia coli*. Alternative hosts, such as Streptomyces, have also been shown to be effective. We can generate new antibiotics by assembling engineered and compatible modules utilizing well-developed hosts as a biological factory.

8.7 BY RIBOSOMAL ENGINEERING

The finding of microorganisms that can tolerate or thrive in high concentrations of organic solvents opens up a new path for producing genetically modified organisms for waste treatment. Therefore, strain improvement is essential to completely utilize the cell's potential. Various scientists observed that a specific ribosomal mutation (a mutation in the rpsL gene encoding the ribosomal protein S12) drastically increased antibiotic development. They concluded that modulating ribosomal proteins or rRNA could dramatically alter bacterial gene expression, leading eventually to the activation of inactive (silent) genes. Through synthesizing a bacterial alarm one, ppGpp, the ribosome plays a unique function in bacterial gene expression. The stringent response, which results in the suppression of stable RNA synthesis in response to nutrient restriction, is one of the most effective adaptation systems for bacteria. The severe reaction is based on a temporary rise in the hyperphosphorylated guanosine nucleotide ppGpp, which is synthesized from GDP and ATP by the relA gene product (ppGpp synthetase) in response to uncharged tRNA binding to the ribosomal A region. Since ppGpp often triggers bacterial secondary metabolism when cells reach the stationary phase, it's critical to consider the stringent response when initiating or enhancing bacterial secondary metabolism.

8.8 STRATEGIES USED TO ENHANCE OVERPRODUCTION OF SECONDARY METABOLITES

Table 8.1 lists the strategies used to enhance secondary metabolite output. These compounds have a significant impact on our society's health, nutrition, and economics. Antibiotics are the most well-known compounds in this category. These extraordinary compounds make up a diverse collection of biologically active molecules with various configurations and modes of action. They attack a wide range of microbial activities, including DNA, RNA, protein synthesis, membrane structure, electron transfer, sporulation, germination, etc.

TABLE 8.1 Strategies Used to Enhance Secondary Metabolite Production

Technique	Metabolites
Metabolic engineering	Anti-tumor agents (anthracenones, anthracycline, glycopeptolides), antibiotics (penicillin G, cephamycin cephalosporin C), xanthan gums, etc.
Protoplast fusion	Penicillin G, cephamycin, cephalosporin C, rifamycin, indolizomycin, etc.
Combinatorial biosynthesis	Erythromycin, tylosin, spiramycins, surfactins, etc.
Whole genome shifting	Spinosad, epothilones
Genome mining	Antibiotics, antifungal compounds, etc.

8.9 CONCLUSION

Due to the use of complementary technologies such as mutation and genetic recombination, many primary and secondary metabolites and protein biopharmaceuticals and enzymes have seen dramatic productivity increases. New genetic approaches to developing overproducing strains are constantly emerging. Metabolic engineering, ribosome engineering, combinatorial biosynthesis, and molecular breeding techniques are effective techniques. Functional genomics, proteomics, and metabolomics are also being used to discover new beneficial molecules for medicine and catalytic enzymes.

KEYWORDS

- **genetic methods**
- **metabolites**
- **mutation**
- **overproduction**
- **ribosomal engineering**
- **secondary metabolites**
- **synthetic biology**

REFERENCES

Adrio, J. L., & Demain, A. L., (2010). Recombinant organisms for production of industrial products. *Bioengineered Bugs, 1*(2), 116–131.

Aharonowitz, Y., & Demain, A. L., (1979). Nitrogen nutrition and regulation of cephalosporin production in *Streptomyces clavuligerus*. *Canadian Journal of Microbiology, 25*(1), 61–67.

Aharonowitz, Y., & Demain, A. L., (1980). Thoughts on secondary metabolism. *Biotechnology and Bioengineering, 22*, 5–9.

Aharonowitz, Y., (1980). Nitrogen metabolite regulation of antibiotic biosynthesis. *Annual Reviews in Microbiology, 34*(1), 209–233.

Baumberg, S., Krügel, H., & Noack, D., (2012). *Genetics and Product Formation in Streptomyces* (Vol. 55). Springer Science & Business Media.

Braña, A. F., Wolfe, S., & Demain, A. L., (1985). Ammonium repression of cephalosporin production by *Streptomyces clavuligerus*. *Canadian Journal of Microbiology, 31*(8), 736–743.

Cimburková, E., Zima, J., Novak, J., & Vanek, Z., (1988). Nitrogen regulation of avermectin biosynthesis in *Streptomyces avermitilis* in a chemically defined medium. *Journal of Basic Microbiology, 28*(8), 491–499.

Corcoran, J. W., (1981). Biochemical mechanisms in the biosynthesis of the erythromycins. In: *Biosynthesis* (pp. 132–174). Springer, Berlin, Heidelberg.

Curdova, E., Kremen, A., Vaněk, Z., & Hostalek, Z., (1976). Regulation and biosynthesis of secondary metabolites. XVIII. Adenylate level and chlorotetracycline production in *Streptomyces aureofaciens*. *Folia Microbiologica, 21*(6), 481–487.

Davati, N., & Najafi, M. B. H., (2013). Overproduction strategies for microbial secondary metabolites: A review. *Int. J. Life Sci. Pharma. Res., 3*, 23–27.

Demain, A. L., & Fang, A., (2000). The natural functions of secondary metabolites. *History of Modern Biotechnology, I*, 1–39.

Demain, A. L., (1971). Overproduction of microbial metabolites and enzymes due to alteration of regulation. In: *Advances in Biochemical Engineering* (Vol. 1, pp. 113–142). Springer, Berlin, Heidelberg.

Demain, A. L., (1998). Induction of microbial secondary metabolism. *Int. Microbiol., 1*(4), 259–264.

Demain, A. L., (1999). Pharmaceutically active secondary metabolites of microorganisms. *Applied Microbiology and Biotechnology, 52*(4), 455–463.

Demain, A. L., (2020). Carbon source regulation of idiolite biosynthesis in actinomycetes. In: *Regulation of Secondary Metabolism in Actinomycetes* (pp. 127–134). CRC Press.

Doull, J. L., & Vining, L. C., (1990). Physiology of antibiotic production in actinomycetes and some underlying control mechanisms. *Biotechnology Advances, 8*(1), 141–158.

Dykstra, K. H., & Wang, H. Y., (1990). Feedback regulation and the intracellular protein profile of *Streptomyces griseus* in a cycloheximide fermentation. *Applied Microbiology and Biotechnology, 34*(2), 191–197.

Fraenkel, G. S., (1959). The raison d'etre of secondary plant substances. *Science,* 1466–1470.

Gonzalez, J. B., Fernandez, F. J., & Tomasini, A., (2003). Microbial secondary metabolites production and strain improvement. *Indian Journal of Biotechnology, 2*, 322–333.

Gräfe, U., (2020). Autoregulatory secondary metabolites from actinomycetes. In: *Regulation of Secondary Metabolism in Actinomycetes* (pp. 75–126). CRC Press.

Horinouchi, S., & Beppu, T., (1992). Autoregulatory factors and communication in actinomycetes. *Annual Review of Microbiology, 46*(1), 377–392.

Hugenholtz, J., & Kleerebezem, M., (1999). Metabolic engineering of lactic acid bacteria: Overview of the approaches and results of pathway rerouting involved in food fermentations. *Current Opinion in Biotechnology, 10*(5), 492–497.

Janglová, Z., Suchý, J., & Vaněk, Z., (1969). Regulation of biosynthesis of secondary metabolites. *Folia Microbiologica, 14*(3), 208–210.

Kominek, L. A., (1975). Cycloheximide production by *Streptomyces griseus*: Control mechanisms of cycloheximide biosynthesis. *Antimicrobial Agents and Chemotherapy, 7*(6), 856–860.

Lebrihi, A., Lefebvre, G., & Germain, P., (1988). Carbon catabolite regulation of cephamycin C and expandase biosynthesis in *Streptomyces clavuligerus*. *Applied Microbiology and Biotechnology, 28*(1), 44–51.

Lee, S. Y., Kim, H. U., Park, J. H., Park, J. M., & Kim, T. Y., (2009). Metabolic engineering of microorganisms: General strategies and drug production. *Drug Discovery Today, 14*(1, 2), 78–88.

Lein, J., (1986). The Panlabs *Penicillin* strain improvement program. *Overproduction of Microbial Metabolites,* 105–139.

Liras, P., Villanueva, J. R., & Martín, J. F., (1977). Sequential expression of macromolecule biosynthesis and candicidin formation in *Streptomyces griseus*. *Microbiology, 102*(2), 269–277.

Lübbe, C., Wolfe, S., & Demain, A. L., (1985). Repression and inhibition of cephalosporin synthetases in *Streptomyces clavuligerus* by inorganic phosphate. *Archives of Microbiology, 140*(4), 317–320.

Magasanik, B., (1961). Catabolite repression. In: *Cold Spring Harbor Symposia on Quantitative Biology* (Vol. 26, pp. 249–256). Cold Spring Harbor Laboratory Press.

Martín, J. F., (1977). Control of antibiotic synthesis by phosphate. In: *Advances in Biochemical Engineering* (Vol. 6, pp. 105–127). Springer, Berlin, Heidelberg.

Martín, J. F., (2020). Molecular mechanisms for the control by phosphate of the biosynthesis of antibiotics and other secondary metabolites. In: *Regulation of Secondary Metabolism in Actinomycetes* (pp. 213–237). CRC Press.

Ochi, K., Okamoto, S., Tozawa, Y., Inaoka, T., Hosaka, T., Xu, J., & Kurosawa, K., (2004). Ribosome engineering and secondary metabolite production. *Advances in Applied Microbiology, 56*(56), 155–179.

Omura, S., & Tanaka, Y., (1984). Control of ammonium ion level in antibiotic fermentation. In: *Biological, Biochemical, and Biomedical Aspects of Actinomycetes* (pp. 367–380). Academic Press.

Omura, S., (1986). Biosynthesis of tylosin and its regulation by ammonium and phosphate. *Regulation of Secondary Metabolite Formation,* 305–332.

Paress, P. S., & Streicher, S. L., (1985). Glutamine synthetase of *Streptomyces cattleya*: Purification and regulation of synthesis. *Microbiology, 131*(8), 1903–1910.

Payne, G. F., & Wang, H. Y., (1989). The effect of feedback regulation and in situ product removal on the conversion of sugar to cycloheximide by *Streptomyces griseus*. *Archives of Microbiology, 151*(4), 331–335.

Pérez-Llarena, F. J., Liras, P., Rodriguez-Garcia, A., & Martin, J. F., (1997). A regulatory gene (ccaR) required for cephamycin and clavulanic acid production in *Streptomyces clavuligerus*: Amplification results in overproduction of both beta-lactam compounds. *Journal of Bacteriology, 179*(6), 2053–2059.

Pospíšil, S., Sedmera, P., Havránek, M., Krumphanzl, V., & Vaněk, Z., (1983). Biosynthesis of monensins A and B. *The Journal of Antibiotics, 36*(5), 617–619.

Raina, S., Murphy, T., Vizio, D., Reffatti, P., & Keshavarz, T., (2011). Novel strategies for over-production of microbial products. *Chem. Eng. Trans., 24*, 847–852.

Shapiro, S., & Vining, L. C., (1983). Nitrogen metabolism and chloramphenicol production in *Streptomyces venezuelae*. *Canadian Journal of Microbiology, 29*(12), 1706–1714.

Shapiro, S., Vining, L. C., Laycock, M., McInnes, A. G., & Walter, J. A., (1985). Pathway of ammonium assimilation in *Streptomyces venezuelae* examined by amino acid analyses and 15N nuclear magnetic resonance spectroscopy. *Canadian Journal of Microbiology, 31*(7), 629–634.

Spížek, J., & Tichý, P., (1995). Some aspects of overproduction of secondary metabolites. *Folia Microbiologica, 40*(1), 43–50.

Spížek, J., Málek, I., Doležilová, L., Vondráček, M., & Vaněk, Z., (1965). Metabolites of *Streptomyces noursei*. *Folia Microbiologica, 10*(5), 259–262.

Spížek, J., Vaněk, Z., & Tichý, P., (1994). Genetic studies on lincomycin production in *Streptomyces lincolnensis*. Proc. 6th European Congr. Biotechnol. In: Alberghina, L., Frontali, L., & Sensi, P., (eds.), *Progress in Biotechnology* (pp. 757–760). Elsevier, Amsterdam.

Stamp, N., (2003). Out of the quagmire of plant defense hypotheses. *The Quarterly Review of Biology, 78*(1), 23–55.

Stein, D., & Cohen, S. N., (1989). A cloned regulatory gene of *Streptomyces lividans* can suppress the pigment deficiency phenotype of different developmental mutants. *Journal of Bacteriology, 171*(4), 2258–2261.

Vanek, Z., & Vondracek, M., (1966). Biogenesis of cycloheximide and of related compounds. *Biogenesis of Cycloheximide and of Related Compounds*, 982–991.

Vaněk, Z., Hošťálek, Z., & Spížek, J., (1990). Overproduction of microbial products-facts and ideas. *Biotechnology Advances, 8*(1), 1–27.

Vaněk, Z., Hošťálek, Z., Blumauerová, M., Mikulík, K., Podojil, M., Běhal, V., & Jechová, V., (1973). The biosynthesis of tetracycline. *Pure and Applied Chemistry, 34*(3, 4), 463–486.

Wittler, R., & Schügerl, K., (1985). Interrelation between penicillin productivity and growth rate. *Applied Microbiology and Biotechnology, 21*(6), 348–355.

CHAPTER 9

Biomass Production

Biomass is among the most available renewable sources of energy on planet Earth. In many developing countries, including India, biomass takes 34 to 40% of the required energy. On the other hand, biomass only accounts for 3–4% of the primary energy source in developed and more industrialized countries. Specifically, the United States attains 3.4% of its energy from biomass, whereas Europe recovers 3.7% of its energy from raw biomass. In countries like Finland and Sweden, 16–18% of electric energy demand is covered by the burning of forest waste. Whereas in Austria.

Interestingly, only 3–4% of biomass energy is utilized in Italy. The distribution considers certain countries' attention and economic effort in developing new techniques and technologies. It concerns the actual availability of land areas where these "energy cultures" are meant to be efficiently implemented. The countries in northern central Europe are among the highly developed countries which use biomass as an energy source. Collective electric energy and biomass-supplied district heating plants are installed in these countries. France has the most significant agricultural area in Europe, and it invests a considerable amount in the production of biodiesel and ethanol. Biomass is the most extensively employed renewable energy source in the developing world, but less so in the developed world. Discovering suitable biomass species that can give high-energy outputs to replace conventional fossil fuel energy sources has been a priority. The energy conversion mechanism and the form in which energy is required decide the type of biomass required.

Fire is without a doubt the most significant discovery in human history. For thousands of years, it has lighted, warmed, guarded, and fed humanity. In a nutshell, fire aided in the development of civilization. On the other hand, the most extensively utilized raw material was Wood for many centuries. It was used as building material along with being used for burning purposes. Mechanical energy was obtained from the combustion of wood with the

Handbook of Industrial Food Microbiology. Manju Nehra & Vikash Nain
© 2025 Apple Academic Press, Inc. Co-published with CRC Press (Taylor & Francis)

invention of the steam engine. The problem of deforestation occurred during the Industrial Revolution, and Wood became scarce. Mankind had to find alternative energy sources, and coal and oil were seen as replacements for Wood. These were at the time abundant but, unfortunately, non-renewable. The pollution problem and the disappearance of fossil fuels led to re-discovering the importance of Wood and biomass as an energy source.

The growth of various microbes such as fungi and bacteria depend upon specific conditions of temperature and humidity. All biomasses are not suitable for multiple biochemical processes such as fungi and bacteria necessarily do not eat everything but only organic matter rich in proteins and water. The sound sources of biomass include seaweeds, rejects from potatoes and beets, food waste, and animal feces. The major products which can be obtained with digestive methods are energy, fertilizers, biogas, bioethanol, and heat. Biogas is a mixture of methane (50–70%) and carbon dioxide (CO_2) that may be utilized for heating. Bioethanol is an alcohol that can be promoted as vehicle fuel. It is produced from the fermentation of beet sugars and sugar cane. This is considered a clean and cheap fuel. It is estimated that 12 million tons of bioethanol are generated each year, particularly in the United States and Brazil.

The decomposition of waste products, including leaves or animal feces, produces heat that can heat greenhouses and other infrastructures like stables. Fuel is the substance used to initiate and sustain a fire. A combustive agent, in addition to fuel, is required for a fire to burn. The most prevalent combustive agent is oxygen, which "oxidizes" the fuel in a combustion reaction, producing light and heat as types of energy. As a result, fire is nothing more than the expression of a chemical interaction, combustion, involving two separate substances, fuel and combustive. There are hundreds of combustible substances and materials found in nature. Initially, Wood was burnt by human being followed by coal. In today's time, the most common fuels used are the fossil fuels such as petroleum, coal, methane gas and or fossilized carbon sources. Modern and efficient heaters that may exploit the hidden significance of Wood and its derivatives have replaced antique fireplaces, chimneys, and stoves. Fungi and bacteria like sticky protein compounds, but fire prefers cellulose-rich dry materials. As cellulose is such a complex molecule, it is extremely resistant. It consists of long glucose chain molecules known as most uncomplicated sugars. The temperature at which wood is burned and the type of physical and chemical transformations that occur are used to classify the many systems that obtain energy from it. Almost all systems use crushed wood. The wood chips can be used as small compact blocks or pellets, increasing the efficiency of most heaters, making them cleaner. Wood transformed into pallets can be combusted at extremely high temperatures (about 1,100°C), and a mixture of gases is formed that can move turbines resulting in producing electric power. Wood is transformed into gases, liquids, and solids when burned at lower temperatures (between 400 and 800°C). Solid fuels, like as coal, can still be used as a source of energy. On the other hand, the liquid component, such as pyrolysis oil, can be used as a fuel for motors as well as a raw material for other products.

9.1 ENERGY PRODUCTION

9.1.1 BIO-POWER

Different processes are employed to extract energy (biopower) from different forms of biomass, and the energy products that result are also diverse. Biopower technologies transform sustainable biomass fuels into heat and electricity using a variety of equipment, similar to how fossil fuels were employed in the past. Biomass has the advantage of being readily available since it may retain its energy until it is utilized in various processes. Thermo-chemical conversion techniques of biomass are based on heat, which stimulates the chemical reactions required to transform the material into energy and can be utilized for cellulose and wood wastes at humidity levels below 30%. Wood and its derivatives, such as sawdust and wood shavings, are the best biomass for thermo-chemical conversion. Cereal straw, grapevine pruning remnants, fruit tree pruning residues, and processing waste such as husks, insect residues, chaff, stones, etc., are the most prevalent wood-cellulose by-products.

9.1.2 BIO-CHEMICAL PROCESSES

Due to the involvement of enzymes, mushrooms, and bacteria, energy is obtained through biochemical reactions. Aquatic breeding, agricultural products/by-products (leaves, sugar beet stems, vegetables, potatoes, etc.), liquid residues of alcohol processing made of grains, vegetation water, and various other types of urban and industrial wastewater are appropriate for chemical conversion.

The techniques dominantly employed in the bio-power sector are: cofiring, pyrolysis, combustion, gasification, small modular systems, anaerobes, and aerobic digestion, and carbonization.

9.1.3 GASIFICATION

At present, as far as efficiency and environmental impact are concerned in terms of biomass, "gasification," a thermo-chemical process, is considered one of the best and most promising technologies to produce electric energy. Each biogas plant/unit is subdivided into various sections, where different stages of the productive process are carried out such as gasification, gas turbine and thermal cycle. During gasification process, the wet biomass is conveyed into a drier making excess water evaporate. After drying the biomass, it is taken to the gasifier, where it is transformed in a mixture of synthetic gas composed of molecular nitrogen (N_2), steam (H_2O), carbon monoxide (CO), CO_2, molecular hydrogen (H_2), methane (CH_4) and a small fraction of much heavier hydrocarbons. Then cooling is done, and synthetic gas is filtered to eliminate dust, contaminating particles (cyanic acid, ammonia, and hydrogen chloride) and various organic compounds (phenols and fatty acids). After that compression is done, it makes the gas turbine ready to operate, where it will be burnt to heat the air that is to be conveyed to the thermal cycle. Coming to the last section of the plant, a boiler recovers the heat in the air coming from the gas turbine. It produces steam for another turbine, operating the electric current generator.

9.1.4 COFING

It is possible to use biomass to complement coal to optimize coal plants. With out a doubt, this is one of the most cost-effective energy alternatives available from renewable sources. The cofiring is based on replacing a portion of coal with biomass located in the already existing plant. It can be done by mixing coal and biomass before the fuel is introduced into the boiler, or it can be done by using separate feeding systems for coal and biomass. Biomass can replace up to 15–18% coal in a cofiring operation, depending on the type of boiler and the feeding mechanism utilized.

9.1.5 PYROLYSIS

Pyrolysis is the thermo-chemical degradation of organic compounds caused by heat application at temperatures between 400 and 800°C in the absence of oxidizing agents. Pyrolysis, like partial gasification, can be done with a much lower amount of oxygen. Pyrolysis products can be gaseous, liquid, or solid, with proportions varying depending on the pyrolysis method (rapid, slow, or conventional) and reaction conditions.

One of the primary issues with pyrolysis-based energy production is the product quality, which has not yet attained an acceptable standard in its applications, whether with gas turbines or diesel engines. Cycles that use pyrolytic oil appear to have a better future, especially for commercial installations.

Diesel engines that employ pyrolysis products, on the other hand, appear to be more efficient. Inside boilers, direct combustion occurs, exchanging heat among combustion gases and process fluids.

9.1.6 COMBUSTION

When compounds like lignin and cellulose, which are rich in structured carbohydrates and have a water content of less than 30–35%, are employed as biofuels, combustion of plant products and agricultural wastes provides good results. Cereal straws, residues of cereals/legumes, debris of oleaginous plants (castor-oil plant, etc.), residues of textile fiber plants (cotton, hemp, etc.), wood in all forms, including wooden residues deriving from the pruning of fruit and forest plants, and residues of the agro-food industry are among the products that can be used in the combustion process.

9.1.7 SMALL-MODULAR SYSTEMS

These technologies have the ability to meet the energy needs of 2.6 billion people who do not have access to electricity. Because most people live in places where huge amounts of biomass can be utilized as fuel, they have this capability. In remote villages, a small system of about 5 megawatts can be an appropriate solution. These compact modular systems are a more environmentally friendly alternative to fossil energy systems. Carbonization is a thermochemical process that converts wood and cellulose products' structured molecules into coal. Carbonization occurs primarily as a result of the removing water and substantial volatile substances from vegetal matter as a result of a heat reaction in a charcoal pile (cone-shaped woodpiles covered in earth, with a central outlet channel (chimney) where slow combustion of wood occurs, converting it into coal. Carbonization happens primarily outside or in flask-shaped containers with long and curved necks, leading to a higher coal output.

9.1.8 *ANAEROBIC DIGESTION*

A biochemical transformation or process that takes place in the absence of oxygen, wherein microorganisms degrade complex organic substances (lipids, carbohydrates, proteins) found in vegetal and animal by-products, producing a gas (biogas) primarily composed of methane (50–70%) and with an average calorific value of 23,000 KJ per cubic meter. Biogas is harvested, dried, compressed, and stored as a result of this process. It can be used to fuel gas boilers to generate heat (and can also be paired with turbines to generate electricity), internal combustion motors (boat engines with a low number of turns are suited for this) or combined-cycle plants. By facilitating organic nitrogen mineralization at the end of the effluent fermentation process, the primary nutritional constituents in the raw material (nitrogen, phosphorus, and potassium) are preserved. The effluent from this method also serves as an excellent fertilizer. Animal excrement, civic trash (wastewater), food waste, and municipal solid waste can all be fed to anaerobic digestion plants with high humidity content leftovers (organic fraction only). Those dumps, on the other hand, are suitable for collecting biogas; nevertheless, only around 40% of the gas produced can be collected. The remainder is diffused into the atmosphere at the same moment. Biogas emissions into the environment are undesirable because methane, which makes up the majority of biogas, is a greenhouse gas with a 20-fold higher effect than CO_2. Almost all of the gas produced by the decomposition of organic waste obtained through anaerobic digestion in (closed) sufficient plant digestors is collected and then used as a fuel. This biogas recovery from landfills is a system that has been evaluated in a lot of nations (England has developed an efficient method of biogas recovery from landfills, both for thermal and electric aims). Biogas refueling stations that supply methane for vehicles are functional in Sweden.

Aerobic digestion metabolizes organic substances through microorganisms, whose development depends on oxygen. According to their metabolic activity, these bacteria convert complex substances into simple ones, releasing CO_2 and H_2O and warming up the substrate. The resulting heat can then be transferred outside through a fluid heat exchanger. In Europe, the aerobic digestion system is used to treat wastewaters. This technology has been spread to Canada and the United States more recently.

9.2 BIOMASS PLANTS

9.2.1 *BIOMASS DISTRICT HEATING*

A district heating system composed of a transportation system and a heat producing facility that serves multiple buildings at the same time.

Cogeneration and/or renewable energy sources can be used in the district heating plant. The heat delivered by urban district heating systems comes from both simple (just heat) and combination (heat + electric energy) generating units. Boilers for heat generation like vapor, hot water, superheated water, and oil are installed at the first step. Combined production installations are cogeneration facilities that can be fed by both a vapor cycle with internal combustion engines and a combined cycle with gas turbines.

The distribution network is the most expensive portion of a district heating plant, accounting for 50–80% of the overall expenditure. Different types of fluids can be used in the distribution system, like in Italy, hot water (between 80 and 90°C) or slightly superheated of 110 to 120°C temperature is used.

9.2.2 DISTRIBUTION SYSTEM

Both direct and indirect distribution systems are used widely. Direct distribution has a single hydraulic circuit that connects the production unit with the user's heating body (radiator or plate), and two separate courses are in contact through a heat exchanger in the indirect distribution system. The direct distribution system needs a lower capital investment and also causes lower heat losses.

9.2.3 DISTRICT HEATING

One immediate solution to explore biomass and for utilization for individual heating in boilers (pellet or nipper) is direct heating with small-sized biomasses (10–12 megawatts). It supplies heat to business and/or household purposes to the sites located near to the biomass production site (Wood, farmland, sawmill, etc.). If biomass is to be produced locally, the plant's dimensions must be carefully considered in order for sources to regenerate. Sizes greater than 10–15 megawatts force an excessive increase in supply, which on the one hand raises the economic and environmental costs of transportation, but on the other hand, going into the wood sector (production, processing, and product sale) is a disadvantage.

More than 300 small rural district heating plants in Austria have biomass, with power ranging from a few 100 kilowatts to 8 megawatts. Plants, on the other hand, are only a few in numbers in Italy, despite the fact that the sector appears to be vital. The main reason for these plants' success is that Austria is linked to an agricultural economy that focuses on forest activities. As a result, there is also a decrease in wood demand, which has resulted in a price drop and problems in the farming sector. This also provides a solution for a reasonable price drop in order to raise the price to an acceptable level for local producers. The majority of district heating plants were built in rural and economically undeveloped areas, but many tourist attractions have also focused on this low-emissions technology. This renewable source appeared to be a component for tourist promotion. In Italy, these plants could also help to alleviate some areas' depression by creating new jobs to maintain forests (a cost-effective and environmentally friendly exercise) and reducing erosion, landslides, floods, and wildfire.

District heating powered by biomass is regarded as a complementary technology. District heating with grill heating systems can incinerate all varieties of wood waste, even if it is very wet and has a low calorific value. Household wood boilers, on the other hand, cannot use this type of garbage, where dry and quality wood of appropriate size was burned. It is necessary to choose pellets to automate the installation in order to avoid constantly supplying it (frequently in winter session). With the same amount of dry mass burned, this Wood takes up less space, ensuring more consistent combustion and easier transport/storage. This variant in the biomass installation supply system allows all products derived from wood waste (branches, barks, roots, etc., especially extremely wet) to be utilized for district heating if sufficient users are available during dry trunks pelletized waste for isolated households. Without a biomass-supplied installation, wood leftovers (from wood processing, agricultural production, sawmills, etc.) would be disposed of in a different way. If they were left out in the open, they would produce the same amount of CO_2 that they had stored during their development. Methane would be produced if fermentation took place in the absence of oxygen. However, if these wastes are disposed of in industries (paper mills, for example), they frequently result in high transportation costs (environmental and economic specially).

The following standards must be met when planning the building of a biomass district heating installation:

- The distance from the supply site should not be excessively lengthy, since transportation can significantly increase the amount of CO_2 generated by the installation as well as the overall cost of the raw material.
- Being close to the supply source allows for a smaller storage capacity inside the district-heating plant (enabling it to be built in smaller spaces), as well as the option of doing the storage at the supplier's location.

- It is vital to have an enough area, close to the transportation network and ideally close to the households, where storage warehouses and installations can be developed without causing too many traffic difficulties.

9.3 BIOFUEL PRODUCTION

The word biofuel refers to the fuels used to produce electric energy, but it generally refers to liquid fuels used for transport. Bioethanol, made from carbohydrates, and biodiesel (ester) derived from fats and oils, are definitely the most frequent fuels. Despite the fact that ethanol is made from starch and sugars, it makes a significant contribution in terms of energy and the environment. Ethanol can also be made from cellulose biomass, such as herbaceous and woody plants, agricultural and forest wastes, and huge quantities of municipal and industrial waste. Starch and sugars represent a moderate quantity of plant material, while cellulose and hemicellulose, polymers of carbohydrate molecules contribute the majority of biomass. Advantage of biofuel is that it has a lower environmental impact than oil derivatives and are made from waste resources that would otherwise go to waste.

Internal combustion engines have always run on ethanol. However, despite their widespread initial availability and inexpensive cost, hydrocarbons were not permitted as fuels. Following the 1973 oil shock, researchers investigated various alternative fuel sources (petrol and gas oil). At the moment, the only product that outperforms ethanol in terms of cost, availability, and performance is ethanol. Four phases are involved in the synthesis of biomass ethanol:

- Biomass generation by the fixation of atmospheric CO_2 to organic carbon.
- The conversion of biomass to a fermentable substrate (often sugar) by one of the numerous technical techniques available is the primary point of distinction between the numerous technical methods for bioethanol conversion.
- Fermentation of intermediate biomass products using bio-catalyzers (yeast and bacteria) to obtain a scarcely concentrated ethanol solution. This stage can be considered the oldest biotechnology ever developed by men.
- By processing the fermentation product.

The result is explosive ethanol and by-products that can be used to produce other fuels, chemical compounds, heat, and electric energy. Even though they are very different, all these last processes conclude with the fermentation synthesis. In alcohol fermentation, the carbohydrates contained in vegetal productions transforms into ethanol. Biodiesel Recycled vegetal oils, kitchen, and animal fats, can be converted to biodiesel by employing a sequence of techniques to initiate and activate these chemical changes at room temperatures that result in the manufacturing of compounds known as esters. It is readily soluble in organic solvents and exist in both solid as well as in liquid state. Then, they are converted into glycerin and biodiesel. Glycerin is a byproduct of the process that can be utilized in the manufacturing of hand creams, lubricants, and toothpaste. Biodiesel is used directly as fuel because it does not require any interference of motors and burners. It may be used for motor propulsion as in diesel engines as pure and in a mixture with the standard gas oil and heating. Biodiesel decreases dependency on oil for energy, greenhouse gas emissions, and health hazards associated with air pollution. Within 30 days, it is non-toxic and biodegradable. Diesel that has been blended with biodiesel has a threefold biodegradability. Biodiesel contains traces of sulfur, in accordance with new EPA (Environmental Protection Agency) criteria, that comes in action from 2006. It is safe to handle and transport except when cold, it may be stored in the same tanks as diesel and pumped using the same equipment (Although tank heater or shaker are necessarily required). It's a robust additive because it can be completely blended with diesel. Biodiesel is an oxygenated substance that aids in the complete combustion of other fuels. Polluting emissions are reduced in proportion to their concentration in mixtures. The emission of NOX is one of biodiesel's drawbacks; experiments are being conducted to address this issue. Due to the varying energy levels, pure biodiesel engines perform 8–15% worse than standard diesel engines. The concerns highlighted earlier are resolved by a 20% combination of diesel and biodiesel. Diesel is a blend of biodiesel (up to 15% by volume) and an additive (to help the two chemicals blend). Spray-mixing is used to make the mixture, which requires no specific equipment or temperature control. When compared to standard diesel, e-diesel effectively reduces particle emissions. Bioproduct: any substance that can be made from fossil fuels may also be made from biomass. As a result, these bioproducts are generated from renewable resources, and they often require less energy for production than their oil-based counterparts. Antifreeze, glue, plastic materials, toothpaste, and synthetic sweeteners may indeed be produced using the same biofuel-making procedures, according to the researchers. Carbon monoxide and hydrogen are two additional reactants used to produce bioproducts. They are synthesized in the presence of oxygen in biomass heating tanks. Biosynthesis gas is a mixture of carbon monoxide and hydrogen that gives life to plastic materials and acids required for the production of photo films, textiles, and synthetic fabrics. Pyrolysis oil is generated when biomass is heated in the absence of oxygen. Phenol is obtained from insulating foam, wooden sticks and plastic molds, as an intermediated product of the process. Methanol, often called wooden alcohol, is typically produced from natural gas, but it can also be made from biomass. The most popular method is biomass gasification, which involves vaporizing biomass at a high temperature, removing hot gas contaminants, and passing it through a catalyzer to convert it to methanol. Corrected petrol compounds derived from biomass can also be used as fuel additives to reduce pollution emissions.

Biomass may be defined as all organic material originating from plants, including algae, shrubs, herbs, trees, and crops. Biomass is mainly produced by green leafy plants which has ability to transform the solar energy into plant food material by mean of photosynthesis and comprises all land and water-based vegetation and all organic residues. The organic matter where the energy of light is retained in the form of chemical bonds can be considered biomass. These substances release their stored chemical energy by breaking down of chemical bonds formed among carbon, hydrogen, and oxygen molecules through different process such as digestion, combustion, or decomposition. Biomass has always been a significant natural energy source for humankind and is presently estimated to contribute to order 10–14% of the world's energy supply.

Plant biomass is used to produce biofuel and is primarily recovered from cellulosic vegetative tissues. Higher growth rate of vegetative plants indicates the higher production of plant biomass. Several molecular and genetic investigations have found variables that control processes like vegetative meristem activity, cell elongation, photosynthetic efficiency, and secondary wall biosynthesis, all of which are important for plant biomass development.

There are many ways to convert plant biomass into energy. The production of gaseous fuel from biomass can be accomplished using a number of different technologies, each with specific requirements, advantages, and disadvantages.

According to several experts, biomass is the plant material produced by photosynthesis, which involves the reaction of CO_2 in the atmosphere, sunlight, and water to build carbohydrates, which are the building blocks of biomass. In most cases, photosynthesis transforms less than 1% of available sunlight into chemical energy. The bonds between the molecules of the building elements of biomass store the sun energy that enables photosynthesis.

The chemical and physical attributes of the macro-molecules from which biomass is formed determine its value. For millennia, man has taken advantage of the energy stored in these chemical bonds as fuel and food for their sugar and starch content. More recently, fossilized biomass has been exploited as coal and oil. Burning fossil fuels use "old" biomass and converts it into "new" CO_2, which contributes to the "greenhouse" effect and depletes a non-renewable resource. The burning of new biomass does not result in a unique CO_2 addition to the environment, as re-planting the harvested biomass guarantees that CO_2 is absorbed and returned to the environment for a new growth cycle to occur. The gap between the instant emission of CO_2 from fossil fuel combustion and the eventual uptake of CO_2 as biomass, which might take years, is often ignored when considering the use of biomass to mitigate global warming. Recognizing this time lag and taking proper action to reduce the length of the lag period are two problems that industrialized countries are grappling with right now. The world is constantly decimating fuel resources while performing poorly to implement a replacement planting program. Several crops have been recommended or are being considered for use as energy sources. Woody crops and grasses/herbaceous plants (all perennial crops), starch and sugar crops, and oilseeds are among the energy crops.

Biomass production had a potential to create huge employment, and if more minor intensively managed energy crops replace intensive agriculture, there are likely to provide environmental benefits, such as reduced leaching of fertilizers with decreased use of pesticides. Compared to current agricultural practice, increased biodiversity depends on the products used and the biomass cultivated. Biomass is renewable, made through natural processes, or human biowaste (organic wastes). The potential biomass energy derived from plant/forest and agricultural residues worldwide is estimated at 30 EJ/yr, compared to the annual worldwide energy demand of over 400 EJ. Suppose biomass contributes to a more considerable extent to the world's energy supply. Further, land that is not suitable for food production can be used for the cultivation of energy farming crops for energy purposes.

Biomass may be converted into three major types of products:

- Electrical/heat energy;
- Transport fuel; and
- Chemical feedstock.

9.4 BIOMASS TYPES

Researchers have characterized various types of biomasses in different ways, but a straightforward method is to define its four different forms as woody plants, herbaceous plants/grasses, aquatic plants, and composts. Within this categorization, herbaceous plants may be further subclassified into plants with high- and low-moisture contents. Numerous industrial operations have now been aimed toward woody plants with lower moisture content. Aquatic plants and manures are relatively intrinsically high-moisture materials and are more suited to 'wet' processing techniques. Biomass with higher moisture level like herbaceous sugarcane plant, lends itself to a 'wet/aqueous' conversion process, such as fermentation.

'Dry' biomass, such as wood chips, on the other hand, is more economically appropriate to gasification, pyrolysis, or combustion. Aqueous processing is generally used when the moisture content of the material is such that the energy required for drying is disproportionate to the energy content of the product formed. Other factors, aside from moisture content, must be taken into account when determining the conversion process's selection, particularly for biomass that falls between the two extremes of 'wet' and 'dry.' Examples include ash, alkali, and trace component contents, which have a negative impact on thermal conversion processes, and cellulose content, which has an impact on biochemical fermentation processes.

Biomass consists of a wide range of cellulose, hemicellulose, lignin, and a few other extractives concentration. Woody plant varieties have slow growth rate and are usually characterized of tightly bound fibers, resulting in a hard outer surface. Herbaceous plants, on the other hand, are usually perennial, with more loosely bound fibers, indicating a lower proportion of lignin, which binds together the cellulosic fibers: both materials are polysaccharides, natural long-chain polymers. The relative amounts of cellulose and lignin define plant species suitability for subsequent processing as energy crops. Cellulose is a glucose linear chain polymer composed of (1,4)-D-glucopyranose units with 1–4 linkage in the beta configuration, with an average molecular weight of around 100,000. Hemicellulose is a polysaccharide combination with an average molecular weight of 30,000 that is mostly composed of sugars such as glucose, mannose, xylose, arabinose, and methyl glucuronic and galacturonic acids.

Hemicellulose is a highly heterogeneous branched polysaccharide that binds firmly but non-covalently to the surface of each cellulose microfibril. Hemicellulose is a five-carbon monosaccharide that differs from cellulose in that it is made up predominantly xylose and other five-carbon monosaccharides. Lignin is an amorphous molecule with a high molecular weight group.

Lignin is made up of three-carbon chains connected to phenyl-propane rings, which have six carbon atoms. These can have zero, one, or two methoxy groups linked to the rings, resulting in three different structures: I, II, and III. The proportion of each structure varies depending on the polymer source, for example, structure I is found in numerous plants such as grasses, structure II in coniferous wood, and structure III in deciduous wood.

Cellulose is generally the most significant portion, contributing around 40–50% of the total biomass by weight; the hemicellulose portion corresponds about 20–40% of total mass.

Photosynthesis is the process by which chlorophyll-containing flora, algae, and some bacteria – capture energy in the form of light and convert it into chemical energy. Photosynthesis provides nearly all of the energy required for life in the biosphere of the Earth. Photosynthesis is divided into two stages: a series of light-dependent, temperature-independent reactions and a sequence of temperature-dependent, light-independent processes. The rate of the light reaction can be accelerated by increasing intensity of the light (within certain limits), but not by raising temperature; this is referred to as the first series. The rate of the dark process can be increased by raising temperature (within certain limits), but not by increasing light intensity, which is referred to as the second series. Another characteristic that distinguishes plant species is the form of photosynthetic pathway that they use. The C_3 photosynthesis method is used by most plants, with the C_3 regulating the amount of carbon in the plant material. Another photosynthetic route in C_4 plants, which accumulates a substantially higher dry mass of carbon than C_3 plants, results in biomass with higher energy conversion potential. Poplar, wheat, willow, and many other cereal crops are C_3 species, whereas perennial grass, Miscanthus, sweet sorghum, maize, and artichoke are C_4 species.

The conversion of biomass into liquids biochemically, e.g., ethanol, has been undertaken using sugar or starch feedstocks, mainly cereals. The process of converting cellulose to glucose through acid/enzymatic hydrolysis and then converting glucose to alcohol. The fermentation method is more easily undertaken with high cellulose content than with biomass rich in lignin, i.e., woody species, which are not readily convertible (Coombs, 1996).

9.4.1 PLANT SPECIES

Plant species selection mainly depends upon the end-use and bio-conversion option of interest, for example, combustion, pyrolysis, fermentation, or mechanical extraction of fats. Almost all possible conversion processes are applicable to some plant species, e.g., oilseed rape can be processed with many methods like combustion, pyrolysis, gasification, mechanical extraction. Wood and cereal crops are suitable only for combustion gasification, pyrolysis, and fermentation. Some plant species may have many benefits for subsequent processing technologies; the amount of energy potentially accessible from a source of biomass remains the same, irrespective of the conversion technology used. However, energy obtained from that biomass may vary between conversion technologies. The attention paid to particular woody and herbaceous plant species varies worldwide, considering soil health and environmental factors that affect growth. When this thing is discussed in the context of northern region of Europe, it is seen that the C_3 woody species are given more attention, particularly poplar and willow, and forest remnants (Ove Arup & Partners, 1989). Herbaceous C_3 species, like cereals, are of less interest in the United Kingdom Because there is little interest in transforming carbohydrate-rich biomass to alcohol (ethanol) fuels, unlike USA and other parts of Europe. Alike, oilseed rape, for producing biodiesel fuel, is of broad interest in Europe but not the UK (Culshaw & Butler, 1992). Brazil is among the pioneers of large-scale fuel–alcohol production derived from sugarcane in tropical climates.

Miscanthus is propagated as rhizomes in which no fertile seed is found, and it is planted in double rows about 70–75 cm apart, with around 170–175 cm distance between rows. Weed inhibition is most important for crop establishment, but the plant faces minor disease problems in Europe. The productivity of 3–10-year-old plants increased in Germany and other countries such as Denmark is 14–32 t/ha: if an output of 20–22 t/ha could be attained, the energy yield would be equivalent to 7–8 t/ha of oil over the duration of each crop. Based on a 20–22 t/ha dry matter yield, the value as an energy crop would be around £622 per hectare, compared to £410 per hectare for wheat; this might correspond to a value one-third that of oil and nearly half that of other sources like coal. Miscanthus production costs are competitive with wood chips from SRC forestry and whole crop cereals. Switchgrass is a C_4 plant with a calorific value similar to that of wood, but with a substantially lower moisture value. When the objective is to produce ethanol for petrol substitution, switchgrass may be a suitable source for conversion processes. It is based on the fact based on the process of saccharification and technology of fermentation with higher ethanol producing capabilities. Ash and alkali content indicates a low value, making it a perfect fuel for combustion, with a shallow potential for slagging. Hemp with an 80% cellulosic content can also be used as a low moisture content biomass feedstock with a high cellulose content (Dewey, 1916). Hemp is a member of the mulberry family (Moraceae), which comprises the mulberry, paper mulberry, and hop plant, and has been cultivated for a long time for a variety of purposes, including medical, nutritional, and chemical manufacturing. It's possible that it'll be utilized to make paper for paper bags. Cannabis hemp seeds are high in essential amino acids and fatty acids, which are necessary for human health.

Linseed oil, on the other hand, is widely used in the production of paints and varnishes, while hemp is the first plant known to have been farmed for the production of a textile fiber. The hemp business in the United States died out in the late 1930s because of its association with marijuana. This relationship inhibits extensive cultivation, despite its potential as a feedstock for a variety of applications.

9.4.2 BIOMASS PROPERTIES

The intrinsic qualities of the biomass source influence both the conversion procedure chosen and any future processing challenges that may occur. Similarly, the type in which the energy is required influences the choice of biomass source. The interaction of these two factors allows for the introduction of flexibility into biomass energy sources. The categories of biomass discussed in this chapter include woody and herbaceous plant species, which are the two types researched by most biomass researchers and technology providers. The material properties affecting subsequent processing as an energy source of biomass may be related to:

- Intrinsic and extrinsic moisture content;
- Calorific value;
- Fixed carbon and volatile proportions;
- Ash content;
- Alkali/metal content;
- Cellulose to lignin ratio.

Former five properties are important for dry biomass conversion processes, whereas the first and last properties are crucial for wet biomass conversion processes.

9.4.3 MOISTURE CONTENT

Like all other products, two types of moisture content are present in biomass:

1. **Intrinsic Moisture:** Moisture that present in the raw material without the influence of external/weather conditions.
2. **Extrinsic Moisture:** This represents the effect of harvesting environmental conditions on the overall moisture content of biomass. Practically, the external moisture content is of concern because the intrinsic moisture content can only be achieved or applied under laboratory conditions.

Putting aside other aspects such as transformation to alcohol or gas/oil, the relationship between moisture content of biomass and bio-conversion technology is essentially straightforward, in that thermal conversion need feedstock with a low moisture content. Bioconversion, on the other hand, can use feedstock with a high moisture content. When thermal conversion technologies are used for feedstocks with high moisture content, the energy balance for the conversion process suffers.

The mineral content of biomass, like sodium, potassium phosphors, calcium, etc., is vital for all thermochemical conversion processes. The reaction of alkali metals with silica in the ash may produce a sticky, mobile liquid phase, clogging airways in the furnace and boiler plant. Contamination with soil introduced during harvesting can significantly increase the total silica content, and the higher concentration of total silica may result in operating complications.

KEYWORDS

- **bio-chemical processes**
- **biofuel**
- **biomass**
- **bio-power**
- **gasification**
- **moisture content**
- **pyrolysis**
- **small-modular systems**

REFERENCES

Abramson, M., Shoseyov, O., & Shani, Z., (2010). Plant cell wall reconstruction toward improved lignocellulosic production and processability. *Plant Sci., 178*, 61–72.

Barratt, D. H. P., Derbyshire, P., Findlay, K., Pike, M., Wellner, N., Lunn, J., Feil, R., et al., (2009). Normal growth of *Arabidopsis* requires cytosolic invertase but not sucrose synthase. *Proc. Natl. Acad. Sci. U S A., 106*, 13124–13129.

Biemelt, S., Tschiersch, H., & Sonnewald, U., (2004). Impact of altered gibberellin metabolism on biomass accumulation, lignin biosynthesis, and photosynthesis in transgenic tobacco plants. *Plant Physiol., 135*, 254–265.

Carpita, N. C., & Gibeaut, D. M., (1993). Structural models of primary cell walls in flowering plants: Consistency of molecular structure with the physical properties of the walls during growth. *Plant J., 3*, 1–30.

Carroll, A., & Somerville, C., (2009). Cellulosic biofuels. *Annu. Rev. Plant Biol., 60*, 165–182.

Coleman, H. D., Yan, J., & Mansfield, S. D., (2009). Sucrose synthase affects carbon partitioning to increase cellulose production and altered cell wall ultrastructure. *Proc. Natl. Acad. Sci. U S A, 106*, 13118–13123.

Demura & Ye, (2010). *Current Opinion in Plant Biology, 13*, 299–304. 303 www.sciencedirect.com (accessed on 18 November 2023).

Demura, T., & Fukuda, H., (2007). Transcriptional regulation in wood formation. *Trends Plant Sci., 12*, 64–70.

Dohleman, F. G., & Long, S. P., (2009). More productive than maize in the Midwest: How does miscanthus do it? *Plant Physiol., 150*, 2104–2115.

Du, J., Mansfield, S. D., & Groover, A. T., (2009). The populus homeobox gene ARBORKNOX2 regulates cell differentiation during secondary growth. *Plant J., 60*, 1000–1014.

Endo, S., Pesquet, E., Yamaguchi, M., Tashiro, G., Sato, M., Toyooka, K., Nishikubo, N., Udagawa-Motosea, M., Kubo, M., Fukuda, H., et al., (2009). Identifying new components participating in the secondary cell wall formation of vessel elements in zinnia and *Arabidopsis*. *Plant Cell, 21*, 1155–1165.

Eriksson, M. E., Israelsson, M., Olsson, O., & Moritz, T., (2000). Increased gibberellin biosynthesis in transgenic trees promotes growth, biomass production and xylem fiber length. *Nat. Biotechnol., 18*, 784–788.

Fujii, S., Hayashi, T., & Mizuno, K., (2010). Sucrose synthase is an integral component of the cellulose synthesis machinery. *Plant Cell Physiol., 51*, 294–301.

Groover, A. T., Mansfield, S. D., DiFazio, S. P., Dupper, G., Fontana, J. R., Millar, R., & Wang, Y., (2006). The populus homeobox gene ARBORKNOX1 reveals overlapping mechanisms regulating the shoot apical meristem and the vascular cambium. *Plant Mol. Biol., 61*, 917–932.

Himmel, M. E., Ding, S. Y., Johnson, D. K., Adney, W. S., Nimlos, M. R., Brady, J. W., & Foust, T. D., (2007). Biomass recalcitrance: Engineering plants and enzymes for biofuels production. *Nature, 315*, 804–807.

Hirano, K., Ueguchi-Tanaka, M., & Matsuoka, M., (2008). GID1-mediated gibberellin signaling in plants. *Trends Plant Sci., 13*, 192–199.

Joshi, C. P., Bhandari, S., Ranjan, P., Kalluri, U. C., Liang, X., Fujino, T., & Samuga, A., (2004). Genomics of cellulose biosynthesis in poplars. *New Phytol., 164*, 53–61.

Ko, J. H., Kim, W. C., & Han, K. H. (2009). Ectopic expression of MYB46 identifies transcriptional regulatory genes involved in secondary wall biosynthesis in Arabidopsis. *The Plant Journal*, *60*(4), 649–665.

Kubo, M., Udagawa, M., Nishikubo, N., Horiguchi, G., Yamaguchi, M., Ito, J., Mimura, T., et al. (2005). Transcription switches for protoxylem and metaxylem vessel formation. *Genes Dev. 19*(16), 1855–1860.

Lee, C., Teng, Q., Huang, W., Zhong, R., & Ye, Z. H., (2009). Down-regulation of PoGT47C expression in poplar results in a reduced glucuronoxylan content and an increased wood digestibility by cellulase. *Plant Cell Physiol., 50*, 1075–1089.

Li, Y., Hagen, G., & Guilfoyle, T. J., (1992). Altered morphology in transgenic tobacco plants that overproduce cytokinins in specific tissues and organs. *Dev. Biol., 153*, 386–395.

Matsumoto-Kitano, M., Kusumoto, T., Tarkowski, P., KinoshitaTsujimura, K., Va′ clavı′kova, K., Miyawaki, K., & Kakimoto, T., (2008). Cytokinins are central regulators of cambial activity. *Proc. Natl. Acad. Sci. U S A, 105*, 20027–20031.

Mauriat, M., & Moritz, T., (2009). Analyses of GA20ox- and GID1-overexpressing aspen suggest that gibberellins play two distinct roles in wood formation. *Plant J., 58*, 989–1003.

McCarthy, R. L., Zhong, R., & Ye, Z. H., (2009). MYB83 is a direct target of SND1 and acts redundantly with MYB46 in the regulation of secondary cell wall biosynthesis in *Arabidopsis*. *Plant Cell Physiol., 50*, 1950–1964.

Melzer, S., Lens, F., Gennen, J., Vanneste, S., Rohde, A., & Beeckman, T., (2008). Flowering-time genes modulate meristem determinacy and growth form in *Arabidopsis thaliana. Nat. Genet., 40*, 1489–1492.

Mitsuda, N., Iwase, A., Yamamoto, H., Yoshida, M., Seki, M., Shinozaki, K., & Ohme-Takagi, M., (2007). NAC transcription factors, NST1 and NST3, are key regulators of the formation of secondary walls in woody tissues of *Arabidopsis*. *Plant Cell, 19*, 270–280.

Morinaka, Y., Sakamoto, T., Inukai, Y., Agetsuma, M., Kitano, H., Ashikari, M., & Matsuoka, M., (2006). Morphological alteration caused by brassino steroid insensitivity increases the biomass and grain production of rice. *Plant Physiol., 141*, 924–931.

Nakano, Y., Nishikubo, N., Goue, N., Yamaguchi, M., Katayama, Y., & Demura, T., (2010). MYB transcription factors orchestrate the developmental program of xylem vessels in *Arabidopsis* roots. *Plant Biotechnol.*

Nieminen, K., Immanen, J., Laxell, M., Kauppinen, L., Tarkowski, P., Dolezal, K., Ta¨ htiharju, S., Elo, A., Decourteix, M., Ljung, K., et al., (2008). Cytokinin signaling regulates cambial development in poplar. *Proc. Natl. Acad. Sci. U S A., 105*, 20032–20037.

Pauly, M., & Keegstra, K., (2010). Plant cell wall polymers as precursors for biofuels. *Curr. Opin. Plant Biol., 13*, 305–312.

Ragauskas, A. J., Williams, C. K., Davison, B. H., Britovsek, G., Cairney, J., Eckert, C. A., Frederick, W. J. Jr., Hallett, J. P., Leak, D. J., Liotta, C. L., et al., (2006). The path forward for biofuels and biomaterials. *Science, 311*, 484–489.

Sakamoto, T., & Matsuoka, M., (2004). Generating high-yielding varieties by genetic manipulation of plant architecture. *Curr. Opin. Biotechnol., 15*, 144–147.

Sakamoto, T., Miura, K., Itoh, H., Tatsumi, T., Ueguchi-Tanaka, M., Ishiyama, K., Kobayashi, M., Agrawal, G. K., Takeda, S., Abe, K., et al., (2004). An overview of gibberellin metabolism enzyme genes and their related mutants in rice. *Plant Physiol., 134*, 1642–1653.

Sakamoto, T., Morinaka, Y., Ohnishi, T., Sunohara, H., Fujioka, S., Ueguchi-Tanaka, M., Mizutani, M., Sakata, K., Takatsuto, S., Yoshida, S., et al., (2006). Erect leaves caused by brassinosteroid deficiency increase biomass production and grain yield in rice. *Nat. Biotechnol., 24*, 105–109.

Salehi, H., Ransom, C. B., Oraby, H. F., Seddighi, Z., & Sticklen, M. B., (2005). Delay in flowering and increase in biomass of transgenic tobacco expressing the *Arabidopsis* floral repressor gene FLOWERING LOCUS C. J. *Plant Physiol., 162*, 711–717.

Scheller, H. V., & Ulvskov, P., (2010). Hemicelluloses. *Annu. Rev. Plant Biol.*

Schrader, J., Nilsson, J., Mellerowicz, E., Berglund, A., Nilsson, P., Hertzberg, M., & Sandberg, G., (2004). A high-resolution transcript profile across the wood-forming meristem of poplar identifies potential regulators of cambial stem cell identity. *Plant Cell, 16*, 2278–2292.

Simmons, B. A., Loque, D., & Ralph, J., (2010). Advances in modifying lignin for enhanced biofuel production. *Curr. Opin. Plant Biol., 13*, 313–320.

Somerville, C., (2006). Cellulose synthesis in higher plants. *Annu. Rev. Cell Dev. Biol., 22*, 53–78.

Somerville, C., (2007). Biofuels. *Curr. Biol., 17*, R115–R119.

Soyano, T., Thitamadee, S., Machida, Y., & Chua, N. H., (2008). Asymmetric leaves2-like/lateral organ boundaries domain genes, ASL19/LBD30 and ASL20/LBD18, control tracheary element differentiation in *Arabidopsis*. *Plant Cell, 20*, 3359–3373.

Sticklen, M. B., (2008). Plant genetic engineering for biofuel production: Towards affordable cellulosic ethanol. *Nat. Rev. Genet., 9*, 433–443.

Takahashi, J., Rudsander, U. J., Hedenstro¨ m, M., Banasiak, A., Harholt, J., Amelot, N., Immerzeel, P., Ryden, P., Endo, S., Ibatullin, F. M., et al., (2009). KORRIGAN1 and its aspen homolog PttCel9A1 decrease cellulose crystallinity in *Arabidopsis* stems. *Plant Cell Physiol., 50*, 1099–1115.

York, W. S., O'Neill, M. A., (2008). Biochemical control of xylan biosynthesis—Which end is up? *Curr. Opin. Plant Biol., 11*, 258–265.

Zhong, R., & Ye, Z. H., (2007). Regulation of cell wall biosynthesis. *Curr. Opin. Plant Biol., 10*, 564–572.

Zhong, R., Demura, T., & Ye, Z. H., (2006). SND1, a NAC domain transcription factor, is a key regulator of secondary wall synthesis in fibers of *Arabidopsis*. *Plant Cell, 18*, 3158–3170.

Zhong, R., Lee, C., & Ye, Z. H., (2010). Functional characterization of poplar wood-associated NAC domain transcription factors. *Plant Physiol., 152*, 1044–1055.

Zhong, R., Lee, C., Zhou, J., McCarthy, R. L., & Ye, Z. H., (2008). A battery of transcription factors involved in the regulation of secondary cell wall biosynthesis in *Arabidopsis*. *Plant Cell, 20*, 2763–2782.

Zhong, R., Richardson, E. A., & Ye, Z. H., (2007). The MYB46 transcription factor is a direct target of SND1 and regulates secondary wall biosynthesis in *Arabidopsis*. *Plant Cell, 19*, 2776–2792.

Zhong, R., Richardson, E. A., & Ye, Z. H., (2007). Two NAC domain transcription factors, SND1 and NST1, function redundantly in regulation of secondary wall synthesis in fibers of *Arabidopsis*. *Planta, 225*, 1603–1611.

Zhou, J., Lee, C., Zhong, R., & Ye, Z. H., (2009). MYB58 and MYB63 are transcriptional activators of the lignin biosynthetic pathway during secondary cell wall formation in *Arabidopsis*. *Plant Cell, 21*, 248–266.

Zhu, X. G., Long, S. P., & Ort, D. R., (2008). What is the maximum efficiency with which photosynthesis can convert solar energy into biomass *Curr. Opin. Biotechnol., 19*, 153–159.

CHAPTER 10

Immobilization of Enzymes

10.1 INTRODUCTION

Enzymes are those specific proteins that speed up various chemical and biochemical transformations. These are biocatalysts found in all organisms (members of plant and animal kingdoms and microorganisms), where they act as catalysts in various vital reactions. The increasing information and technical advances regarding protein extraction and purification techniques opened the door to producing several enzymes with analytical grade purity for research and other technological uses. Enzymes are used as eco-friendly biocatalysts in many industries such as chemicals, pharmaceuticals, and food industries for commercial production and as biosensors for target analysts' identification or determination. However, their biological origin produces hindrance in their commercial applications such as enzymes solubility, temperate stability, feed-back inhibition either by substrates, products, or any other substance, etc.

In comparison to free enzymes, immobilized enzymes possess a higher potential and tolerate the environmental conditions used in chemical reactions. Further, their heterogeneity makes it easy to recover enzymes and products and allows multiple uses of the same enzymes. It also helps maintain continuous and quick reaction termination processes and better control over them with higher variation freedom of bioreactor designs. Immobilization emerged as a powerful solution for the problem associated with the soluble nature of enzymes in the reaction system. Enzyme immobilization is a combination of the enzymes and carrier with maintaining selectivity, stability, and kinetics of enzymes along with chemical and physical characteristics of the carrier substance in a particular formulation in such a way that has a primary role in maximizing the stability of the biocatalyst.

Handbook of Industrial Food Microbiology. Manju Nehra & Vikash Nain
© 2025 Apple Academic Press, Inc. Co-published with CRC Press (Taylor & Francis)

In this method, enzyme immobilization has emerged as a potential instrument in biocatalyst design, accompanying former methods usually employed to increase protein characteristics, such as genetic, microbiological, or chemical modification tools.

10.2 IMMOBILIZATION STRATEGIES

Although the enzyme has many advantages over conventional catalysts, there are several practical difficulties related to its use on a commercial scale. They are usually expensive; that is isolation and purification cost of enzymes is many folds higher than conventional catalysts. Enzymes being protein in the structure are susceptible to various denaturing conditions when isolated from their natural habitats. Different sensitive parameters, like temperature, pH, and micro-level substances, can act as inhibitors that enhance their costs.

While on the other hand, unlike chemical catalysts, enzymes are dissolved in the water and are very difficult to recover in active form after completion of reaction for reuse purposes. The most successful method to overcome these limitations is the Immobilization of enzymes. It is a technical process where enzymes and carriers (also known as solid support) are fixed, producing a diverse immobilized enzyme organization. Immobilized enzymes mimic like they behave naturally in living cells, where enzymes are mostly found attached to cell membrane, organelle, and cytoskeleton. Being more sustainable and resistant to environmental conditions, diversified immobilized enzyme system make is easy to recover product and enzymes, repeated enzymes, fast and controlled reaction termination, continuous operation of enzymatic processes and bioreactor designs. Contrarily, they have low activity and high apparent Michaelis constants compared to free enzymes due to relatively poor substrate accessibility.

10.2.1 MODES OF IMMOBILIZATION

Various methods are used for enzyme immobilization, but industry always prefers cost-effective and straightforward methods. The most commonly employed techniques are based on physical and chemical enzyme immobilization. Five principal methods are used for enzyme immobilization. These are adsorption, encapsulation, entrapment, covalent binding and cross-linking. This binding can be physical or chemical support, involving weak or covalent bonds. Physical bonding is weak compared to chemical bonding and is rarely able withstand against the industrial conditions to keep the enzyme fixed. The support used in this process is usually a biopolymer or synthetic resin, also known as an inorganic polymer such as zeolite or silica.

Entrapment is a method of including an enzyme in a polymeric network (gel lattice) such as a silica-based sol-gel, microcapsule, organic polymer or membrane device. It requires the production of the polymeric system in the presence of the enzyme. The final group includes using a bi-functional reagent to cross-link enzyme aggregates or crystals to create carrier-free macro-particles. Numerous other approaches have been created, some of which are novel variations of the ones mentioned, while others are unique to a particular support or enzyme. Although every approach has certain benefits and limits (Figure 10.1).

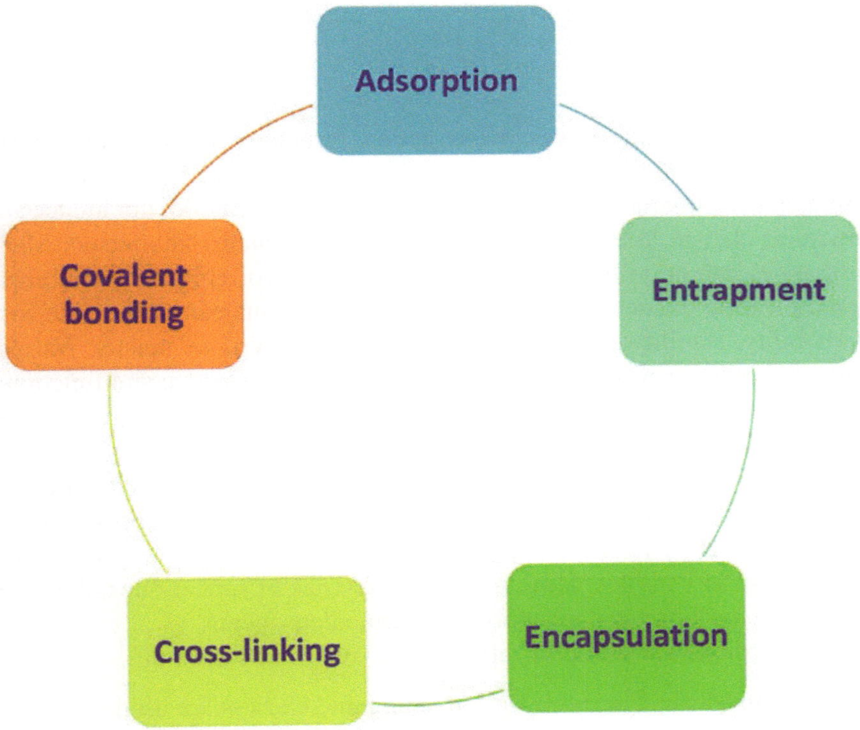

FIGURE 10.1 Various methods of immobilization.

10.2.1.1 ADSORPTION

Among various available methods, the most applied method is adsorption where different enzyme-carriers to fix on solid are used to immobilize enzymes. This term was supposed to be first used in the early 20th century and directly denoted to the enzymes constrained by carriers. These days, it is employed for both direct and intermediate agent-mediated enzymes' Immobilization on the carriers. Various chemical and physical techniques are used for the enzyme immobilization on the solid carriers. Among these suggested approaches, activating enzymes by adsorption is the most effective and suitable method. This is most simple, cheap, and easy method of preparing an immobilized enzyme. Various materials used in this technique are activated charcoal, alumina, ion exchange resins, etc. The only disadvantage associated with is poor association of enzyme and carrier. This process is classified as a carrier-bound immobilization technique, which is reversible. Adsorbent is the simplest and oldest method of Immobilization.

10.2.1.1.1 Principle

This method uses the physical interactions produced between the carrier and enzyme, including van der Waals forces, hydrogen bonding, and ionic interactions. Although binding in this method is week, it does not induce any structural change in enzyme and hence activity remains unaffected. Many substances can be effectively used as the carriers for enzyme immobilization. Selection of such compounds is based on various parameters like cost, availability, stability in specified conditions and type of reactor. Various physical parameters like surface area, pore size, particle size and functional group are considered. It is important to understand that every carrier can be used for enzyme adsorption, but every enzyme cannot be immobilized on all carriers. This is because successful adsorption requires certain conditions to be fulfilled, including enzyme-carrier affinity. Presences of specific active functional groups fulfill this condition and make is suitable for forming enzyme carrier complex. However, in the absence of such groups, the interactions can be generated by applying intermediate agents.

Carriers used for Immobilization of enzymes by adsorption:

1. **Inorganic Carriers:** Various inorganic carriers are used in adsorption-mediated enzyme immobilization, but silica gained particular attention. Silica has typical porous morphological structures and varying dispersive properties, making it suitable for carriers. Its small pore size (5–30 nm diameter) and a hexagonal array of pores make it unique. Some other inorganic carriers used in this process are zirconia, alumina gel, aluminum, and titania sol-gel film.

It is also reported that mullite and cordierite are used for enzyme immobilization. The first one (mullite) is an aluminosilicate-based mineral, while the latter (Cordierite) is a rare mineral associated with the silicate group. Some other minerals are also reported to act as carriers in enzyme immobilization, such as halloysite, mica, hydroxyapatite, and bentonite.

2. **Organic Carriers:** Out of various organic carriers, chitosan, and alginate are of specific interest for enzyme adsorption. A polymer of chitin, also known as Chitosan, is obtained from chitin by deacetylation. Its ability to produce various geometric configurations enables it to be used in a diversified manner like beads, fibers (both sponge and hollow), nanoparticles, and membranes. Further, its non-toxic, gel-forming ability with cationic substances and biocompatibility make it unique for this purpose. Inorganic substance, Calcium alginate is another compound that is applied as a carrier for enzymes. An anionic polysaccharide, alginate has excellent gel-forming behavior with thickening and stabilizing properties. Alternatively used, the organic carrier is cellulose. This is a natural polysaccharide made of glucose molecules. Agarose gel, a polysaccharide polymer, is also found applied in the Immobilization of enzymes because of its attractive adsorption abilities and unique morphological characteristics.

Apart from these polymers from natural origin, a large and diverse cluster of synthetic polymers are carriers for various enzymes. The most usually employed artificial polymers are PVA (polyvinyl alcohol), poly methylol, acrylamide, and polypropylene. Hydroxybutyrate is a unique carrier for enzyme immobilization with biodegradable and thermo-shrinkable characteristics and is used in its nanoparticles form. Furthermore, poly acrylonitrile (PAN), a polymer commonly used to produce synthetic fibers, can also be an excellent carrier in enzyme immobilization. It is elementary and inexpensive material with stiff and elastic nature having high porosity and ability to interact with other materials. Some other substance such as artificial oil bodies and olive pomace powder are also have excellent adsorption properties and used for the enzyme immobilization.

10.2.1.1.2 Surface Modifying Agents

The primary requirement of adsorption immobilization is presence of particular functional groups on surface of both the components that participate in this process, i.e., enzymes and the carrier. These provide sufficient interactions between enzyme and carriers for binding. In their absence, carriers are chemically modified to generate such properties. For this purpose, chemical used are known as surface modifying agents. These agents must have two functional groups: one that allows them to anchor chemically on the carrier substance and another that allows them to interact physically with the enzymes.

Glutaraldehyde, widely use as disinfectant and preservative, have active aldehyde groups and can act as an enzyme immobilizer because of its strong affinity for proteins, fungi, and bacteria. Polyethyleneimine, polystyrene, and poly styrene sulfonate are widely used surface modifying agents for enzyme immobilization. Some amphoteric nature chemicals with excellent hydrogen bond formation properties are also used for this purpose. Some common examples are diethylamine, diethylaminoethyl (DEAE) and monoaminomethyl-N-aminoethyl. Carboxylic acids also have such properties and widely used as modifying agent. Erucic acid and itaconic acid are examples of this category successfully used for enzyme immobilization.

10.2.1.2 COVALENT BONDING

Covalent binding is a conventional technique of enzyme immobilization that involves the direct association of the enzyme to the carrier substance utilizing covalent bonding. This is firm and durable binding enzymes with support material and various substances act as support material such as agarose, porous glass and silica, polyacrylamide, etc. A selection criterion of covalent method of Immobilization is that a reaction process does not require enzyme in the product. This method has the expected advantage of strong interactions between the enzyme and the support materials during the catalytic process that provide a high stability to enzymes.

10.2.1.2.1 Principle

The support material is activated by production of electrophilic group on it which help in coupling the carrier substances with the strong nucleophiles present on the enzymes. Covalent bonds are usually developed between the active functional group of support materials and the amino acid residues on the enzyme surface. The sulfhydryl group of cysteine, the hydroxyl group of threonine and serine, are the major amino acid residues that participate in the covalent attachment of enzymes and carriers. Direct association or the spacer arms are usually employed to bind the enzyme and the support materials. The advantage of the spacer arm is that it allows the enzymes to move around more freely, resulting in higher enzyme activity than a direct association.

10.2.1.3 ENTRAPMENT

In contrast to adsorption and covalent bonding, the Immobilization by entrapment allows enzyme molecules to move freely in solution but constrains their movement in the lattice arrangement of a gel. Here, enzymes are immobilized through their entrapment in a carrier or within fibers.

10.2.1.3.1 Principle

Since enzymes are huge macromolecules, they appear to be greater than the substrates or materials they interact with. As a result, the enzyme is immobilized inside a matrix with a particular pore size that permits only substrates and products with pore sizes smaller than the matrix pore size to diffuse in and out of the mesh. The size of the enzyme and pore is the determining factor for selecting support. Smaller pore size decreases the enzyme entrapment, while larger pore size allows the higher leakage of the enzyme. Hence, appropriate pore size selection of the support is crucial.

A two-stepped process carries out the immobilization:

- Enzymes are mixed into a monomeric solution, followed by
- A chemical reaction or modifying experimental conditions promotes the polymerization of this monomeric solution.

In this process, as an enzyme is confined physically within a polymer lattice network, no chemical interaction occurs between enzyme and entrapping polymer. Hence, this method has better enzyme stability and minimum leaching and denaturation of the enzyme. Further, it is also possible to optimize the microenvironment for the enzyme by altering the entrapping material.

However, since polymerization extension raises the gel matrix thickness, the substrate cannot diffuse far enough into the gel matrix to enter the enzyme active site, which is a major disadvantage associated with this process. Some entrapping strategy is adopted widely depending upon the type of entrapment involved. These include microencapsulation, photo-polymerization, electro-polymerization, and sol-gel method.

10.2.1.4 CROSSLINKING

Crosslinking is an irreversible technique used for enzyme immobilization by means of producing inter-molecular cross-linked bonds among enzymes through covalent bonding. This process is performed with the aid of multi-functional agents that can bind with multiple enzymes simultaneously in a three-dimensional way to form an aggregate of enzymes. Here enzyme does not bind to any support materials and remains present in the reaction mixture. Usually, two different immobilization methodologies by cross-linking are used: cross-linking enzyme aggregate (CLEA) and cross-linking enzyme crystal (CLEC).

10.2.1.4.1 Cross-Linking Enzyme Crystals (CLEC)

The enzyme is crystallized using glutaraldehyde in this step. As a result, the cross-linking agent is used to obtain enzyme crystals. When added to the reaction mixture, the CLEC catalyzes the reaction with a high-efficiency level. High-temperature or organic chemicals do not denature the CLEC, making them very stable. They have a moderate level of proteolysis resistance. Their size is very much controllable with high stability in bioreactor during operation and can easily be recycled. They often improve the enantioselectivity of specific substrate forms.

10.2.1.4.2 Cross-Linking Enzyme Aggregates (CLEA)

Cross-linking enzyme aggregate (CLEA) is an advanced form of CLEC processing that eliminates CLEC's drawbacks. CLEA could act in aqueous solutions, while CLEC involves the creation of crystals. When salts, organic solvents, or non-ionic polymers are added to enzymes, aggregates develop that maintain catalytic activity. These aggregates are known as Cross Linked Enzyme Aggregates (CLEA).

- **Advantage:** It is less expensive, simpler to perform, and has a broad variety of applications. This technique is utilized to synthesize the multi-enzyme catalyst having enzyme in aggregate also known as combi-CLEA. The enzyme may be cross-linked with other enzymes or with a chemically inactive protein, e.g., bovine serum albumin. The simplicity of this approach and the solid chemical bonds formed between biomolecules make it attractive. CLEAs are focused on multi-point association between enzyme molecules under intermolecular cross-linking agents. Various investigators have evaluated the effective development of CLEAs from a wide variety of enzymes, such as lipases, horseradish peroxidase laccases and penicillin acylases, etc.

- **Disadvantage:** The disadvantage of this method is the potential enzyme activity loss due to structural and chemical alteration in the enzyme's active site on cross-linking.

10.2.1.4.3 CROSSLINKING REAGENTS

Glutaraldehyde is the most commonly used and preferred cross-linking reagent for most enzymes due to its low cost, high stability, and easy availability in large amounts for industrial purposes. Although it has some negative performance with certain enzymes and restricts or completely blocks their activity, there are specific alternative available to overcome these limitations. Some of them are bis-isocyanate, diazonium salts, dextran polysaccharide bis-diazobenzidine and some inactive proteins like bovine serum albumin.

10.2.1.5 ENCAPSULATION

Encapsulation is the process of putting nutritional additives, enzymes, cells, or other materials into tiny capsules. It is specified technique of entrapping an ingredient/cell/enzyme in a semi permeable membrane. This technique has gained a significant interest due to its ability to mask the effect of moisture, heat or other extreme conditions on targeted component by using suitable encapsulating material.

10.2.1.5.1 Principle

Its fundamental theory is premised on the fact that the enzyme is more significant than the substrates or compounds it interacts with. Therefore, enzymes are entangled in a semipermeable membrane, allowing the passage of substrate and product while preventing the enzyme from going through it. The pore size of membrane restricts the passage of enzymes and thus retains it within the membrane/capsule. Membrane permeability can be regulated based on the size and form of immobilized enzyme.

Enzyme encapsulation can be accomplished using one of two methods:

1. Coacervation: By using a water-immiscible solvent, the polymer will split enzyme micro-droplets.

2. Interfacial Polymerization: Here an aqueous enzyme solution is mixed with a hydrophobic monomer and second monomer is already dispersed in immiscible solvent. This encourages the process of polymerization on the interface of droplets and, thus, encapsulates the enzyme.

   - **Advantages:** This immobilization method keeps the enzyme molecule in its natural state and eliminates the chances of leakage; it also protects the enzyme against the reaction parameter used. Multi-enzyme encapsulations, which can also be formed by entrapping different types of enzymes inside a membrane, are another benefit. Although, identical dimension of enzyme and product limit encapsulation applications in the immobilization process (Figure 10.2).

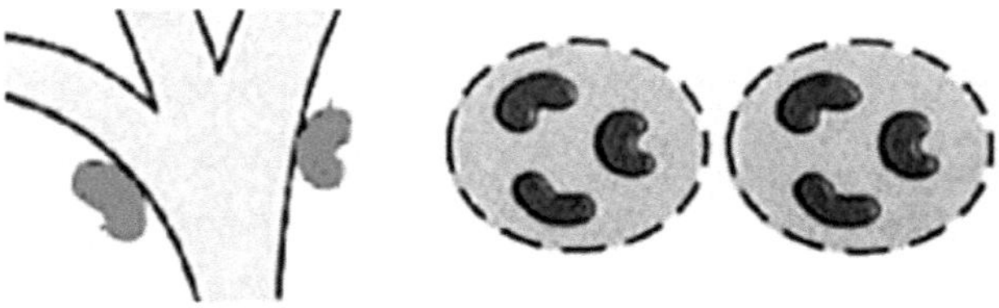

FIGURE 10.2 Encapsulation immobilization.
Source: Modified from Homaei et al. (2013)

10.3 APPLICATIONS OF IMMOBILIZED ENZYMES

Enzyme immobilization is a frequently used approach in various fields, including food and non-food industries such as dairy, pharmaceuticals, bioremediation, detergents, textiles, etc. This approach is used because of its technical and financial benefits. Various enzymes have been immobilized in large quantities and widely used in various large-scale applications.

10.3.1 FOOD INDUSTRY

Immobilized enzymes perform a critical function in the food sector and have diversified applications in different food industries. It is essential to manufacture food products on a large scale with a cost-effective processing approach in the food sector. Hence, the biocatalyst (immobilized enzyme) must be inexpensive in such conditions, because it must have operated solid reliability and complete a significant number of cycles. The food industry often desires continuous flow operational arrangements over batch processes, mainly when large quantities of products are produced. Production of high fructose corn syrup (HFCS), galacto oligosaccharides, and cocoa butter analogs are some common examples of immobilized enzymes application on large scale in continuous process for the manufacturing of foodstuffs (Table 10.1).

TABLE 10.1 Applications of Some Immobilized Enzymes in the Food Industry

SL. No.	Name of the Enzyme	Substrate	Product
1.	Aminoacylase	Amino acids	L-Amino acids
2.	β-Galactosidase	Lactose	Lactose-free milk
3.	Glucoamylase	Starch	D-Glucose
4.	Invertase	Sucrose	Glucose and fructose
5.	Lipase	Triglycerides	Cocoa butter substitutes
6.	Papain	Proteins	Removal of 'chill haze' in beers
7.	Thermolysin	Peptides	Aspartame

1. **Alcoholic Beverage Industry:** The use of immobilized yeast cells allows for a higher concentration of cells in a particular area of the fermenter. It enables the cells to be held and facilitates the down-stream processing. For the manufacturing of alcohol, immobilized cells are used industrially. Mainly the fermentation process is carried out in a bioreactor with immobilized alcohol-producing bacterial cells. It has been reported that immobilized cells were found to have a higher production rate than those of non-immobilized cells. High production rate and easy product recovery have encouraged their industrial application in the brewing sector.
2. **Production of HFSC:** In terms of the amount of enzyme utilized and the volume of product generated, the application of immobilized D-Glucose/xylose isomerase in manufacturing HFCS is the main industrialized procedure that involved an immobilized enzyme. In addition, over 500 tons of immobilized Glucose/xylose isomerase are used per year, allowing approximately 10 million tons of HFCS to be produced annually. HFCS is used primarily to manufacture fructose, which is further used as a sweetener in beverages and other food industries as a food ingredient.
3. **Dairy Industry:** In the dairy industry, immobilized enzymes are frequently used to obtain the starting components and enable various reactions, such as cheese and yogurt processing, and prevent the growth of psychotropic bacteria in milk. Immobilization of enzymes can also affect lactose and citrate metabolism, allowing dairy products to be nutritionally enhanced. Applications of immobilized lipases

have enabled the addition of linolenic acid (an anti-carcinogen) in dairy products.

4. **Meat Industry:** Cell immobilization may somehow benefit the meat industry by decreasing starter culture costs. It also ensures equal and systematic distribution of bacteria, which simplifies the recovery and recycling procedures. Immobilizing cells contribute to meat fermentation. The solid and semisolid nature of meat substrate restricts cell mobility. When immobilized cells are used in meat starter cultures, it makes the production of a continuous cell line possible, ensuring the retention of stable cell lines.
5. **Cider Formation:** Despite the conventional feature of cider-producing industry research, the transformation of apple juice into cider is a complicated method involving the activity of yeast and lactic acid bacteria (LAB) to achieve respective alcoholic and malolactic fermentations. For the improvement in process control of these two fermentations and the production of flavor developing compounds, great interest has been seen in the use of cell immobilization technology to cider production in last two decades.
6. **Production of Enzyme:** Microorganisms are the preferred sources of enzymes. But some rare enzymes are only existed in plant sources such those found in medicinal plants. Plant cell immobilization techniques are applied for effective and large-scale production of such enzymes.
7. **Antibiotics:** Cell cultures are not used in traditional antibiotic manufacturing processes. Various attempts have been made to produce different secondary metabolites by applying stressed conditions cell incubation technique. Since the immobilized cells are restricted to a fixed zone, it is easier to design the stressful conditions needed for secondary metabolite production. Application of immobilized β-acylase has shown higher efficiency in Penicillin G hydrolysis. Similarly, enzymatic production of Cephalexin by applying immobilized Penicillin G Acylase also found promising results.
8. **Wastewater Treatment:** Currently wastewater treatment is becoming necessity due to the growing consumption of fresh water and mixing of the untreated industrial wastewater in water bodies. Textile, paper, and leather industries are both sources of dye effluents, which contain high concentrations of dye in these effluents.

Even a low concentration of these effluents poses a threat to the environment; they are carcinogenic also. These days, enzymes are applied for the degradation of such dyes. Mainly used enzymes in this process are laccase, peroxidases, and azo-reductases. Extreme temperatures, low to high pH, and high ionic strength may cause activity loss in these enzymes; to solve this issue, immobilized enzymes are used these days. The enzyme is covered by a nanometer thick material in Single enzyme nanoparticle since it has a wide surface region. These complexes tend to retain their enzymatic activity in even highly adverse conditions and hence found their application in wastewater treatment specifically in removing heavy metals.

KEYWORDS

- **covalent bonding**
- **cross-linking enzyme aggregates**
- **cross-linking enzyme crystals**
- **crosslinking reagents**
- **encapsulation**
- **immobilization**
- **surface modifying agents**

REFERENCES

Abdelmajeed, N. A., Khelil, O. A., & Danial, E. N., (2012). Immobilization technology for enhancing bio-products industry. *African Journal of Biotechnology, 11*(71), 13528–13539.

Aggarwal, S., & Sahni, S., (2012). The commercial exploitation of immobilized enzymes. *International Proceedings of Chemical, Biological and Environmental Engineering, 41*, 18–22.

Ajitha, S., & Sugunan, S., (2010). Tuning mesoporous molecular sieve SBA-15 for the immobilization of α-amylase. *Journal of Porous Materials, 17*(3), 341–349.

Basso, A., & Serban, S., (2019). Industrial applications of immobilized enzymes—A review. *Molecular Catalysis, 479*, 110607.

Bautista, L. F., Morales, G., & Sanz, R., (2010). Immobilization strategies for laccase from *Trametes versicolor* on mesostructured silica materials and the application to the degradation of naphthalene. *Bioresource Technology, 101*(22), 8541–8548.

Bayramoglu, G., Gursel, I., Yilmaz, M., & Arica, M. Y., (2012). Immobilization of laccase on itaconic acid grafted and Cu (II) ion chelated chitosan membrane for bioremediation of hazardous materials. *Journal of Chemical Technology and Biotechnology, 87*(4), 530–539.

Becker, W., (2004). 21 microalgae for aquaculture. *Handbook of Microalgal Culture: Biotechnology and Applied Phycology,* 380.

Bellino, M. G., Regazzoni, A. E., & Soler-Illia, G. J., (2010). Amylase-functionalized mesoporous silica thin films as robust biocatalyst platforms. *ACS Applied Materials and Interfaces, 2*(2), 360–365.

Berna, B. M., & Batista, F., (2006). *Enzyme Immobilization Literature Survey Methods in Biotechnology: Immobilization of Enzymes and Cells.* 10–15.

Bernal, C., Rodriguez, K., & Martinez, R., (2018). Integrating enzyme immobilization and protein engineering: An alternative path for the development of novel and improved industrial biocatalysts. *Biotechnology Advances, 36*(5), 1470–1480.

Brady, D., & Jordaan, J., (2009). Advances in enzyme immobilization. *Biotechnology Letters, 31*(11), 1639.

Bronshteyn, M., & Steinberg, D., (2002). Immobilization of fructosyltransferase from *Streptococcus* mutans on hydroxyapatite surfaces induces the formation of multimeric complexes. *Letters in Applied Microbiology, 34*(3), 205–209.

Cantone, S., Ferrario, V., Corici, L., Ebert, C., Fattor, D., Spizzo, P., & Gardossi, L., (2013). Efficient immobilization of industrial biocatalysts: Criteria and constraints for the selection of organic polymeric carriers and immobilization methods. *Chemical Society Reviews, 42*(15), 6262–6276.

Cao, L., (2005). Immobilized enzymes: Science or art? *Current Opinion in Chemical Biology, 9*(2), 217–226.

Cao, L., Van, L. L., & Sheldon, R. A., (2003). Immobilized enzymes: Carrier-bound or carrier-free? *Current Opinion in Biotechnology, 14*(4), 387–394.

Castro, L. B., Silva, F. F., Carmona-Ribeiro, A. M., Kappl, M., & Petri, D. F., (2007). Immobilization of hexokinase onto chitosan decorated particles. *The Journal of Physical Chemistry B, 111*(29), 8520–8526.

Chen, Y. Z., Yang, C. T., Ching, C. B., & Xu, R., (2008). Immobilization of lipases on hydrophobilized zirconia nanoparticles: Highly enantioselective and reusable biocatalysts. *Langmuir, 24*(16), 8877–8884.

Crabb, W. D., & Shetty, J. K., (1999). Commodity scale production of sugars from starches. *Current Opinion in Microbiology, 2*(3), 252–256.

De Lathouder, K. M., Van, B. D. T. J., Wallin, S. A., Mateo, C., Lafuente, R. F., Guisan, J. M., & Moulijn, J. A., (2008). Polyethyleneimine (PEI) functionalized ceramic monoliths as enzyme carriers: Preparation and performance. *Journal of Molecular Catalysis B: Enzymatic, 50*(1), 20–27.

Deepak, V., Kalishwaralal, K., & Gurunathan, S., (2009). Purification, immobilization, and characterization of nattokinase on PHB nanoparticles. *Bioresource Technology, 100*(24), 6644–6646.

Durieux, A., Nicolay, X., & Simon, J. P., (2005). Application of immobilization technology to cider production: A review. In: *Applications of Cell Immobilization Biotechnology* (pp. 275–284). Springer, Dordrecht.

Erhardt, F. A., & Jördening, H. J., (2007). Immobilization of dextranase from *Chaetomium erraticum*. *Journal of Biotechnology, 131*(4), 440–447.

Fornera, S., Bauer, T., Schlüter, A. D., & Walde, P., (2012). Simple enzyme immobilization inside glass tubes for enzymatic cascade reactions. *Journal of Materials Chemistry, 22*(2), 502–511.

Garcia, H. S., Storkson, J. M., Pariza, M. W., & Charles, Jr. G., (1998). Enrichment of butteroil with conjugated linoleic acid via enzymatic interesterification (acidolysis) reactions. *Biotechnology Letters, 20*(4), 393–395.

Ghous, T. N., (2001). Analytical application of immobilized enzymes. *Journal-Chemical Society of Pakistan, 23*(4), 228–233.

Górecka, E., & Jastrzębska, M., (2011). Immobilization techniques and biopolymer carriers. *Biotechnology and Food Science, 75*(1), 65–86.

Grudzien, R. M., Grabicka, B. E., Felix, R., & Jaroniec, M., (2007). Polymer-templated organosilicas with hexagonally ordered mesopores: The effect of organosilane addition at different synthesis stages. *Adsorption, 13*(3, 4), 323–329.

Gupta, A., & Prabhu, K. A., (1995). Immobilization and properties of dextransucrase from *Leuconostoc mesenteroides* culture, LM1. *The Journal of General and Applied Microbiology, 41*(5), 399–407.

Hanefeld, U., Gardossi, L., & Magner, E., (2009). Understanding enzyme immobilization. *Chemical Society Reviews, 38*(2), 453–468.

Hartmann, M., & Kostrov, X., (2013). Immobilization of enzymes on porous silicas–benefits and challenges. *Chemical Society Reviews, 42*(15), 6277–6289.

Hasan, F., Shah, A. A., & Hameed, A., (2006). Industrial applications of microbial lipases. *Enzyme and Microbial Technology, 39*(2), 235–251.

Hemalatha, V., Kalyani, P., Chandana, V. K., & Hemalatha, K. P. J., (2016). Methods, applications of immobilized enzymes-a mini review *International Journal of Engineering Sciences and Research Technology, 5*(11), 523–526.

Homaei, A. A., Sariri, R., Vianello, F., & Stevanato, R., (2013). Enzyme immobilization: An update. *Journal of Chemical Biology, 6*(4), 185–205.

Hua, Y. W., Fang, M., Shen, T. D., Shao, P., Ning, X. T., & Hui, Z. C., (2013). Immobilization of *Candida rugosa* lipase on hexagonal mesoporous silicas and selective esterification in nonaqueous medium. *Biochemical Engineering Journal, 70*, 97–105.

Hung, Y. J., Peng, C. C., Tzen, J. T., Chen, M. J., & Liu, J. R., (2008). Immobilization of *Neocallimastix patriciarum* xylanase on artificial oil bodies and statistical optimization of enzyme activity. *Bioresource Technology, 99*(18), 8662–8666.

Hwang, E. T., & Gu, M. B., (2013). Enzyme stabilization by nano/microsized hybrid materials. *Engineering in Life Sciences, 13*(1), 49–61.

Jesionowski, T., Zdarta, J., & Krajewska, B., (2014). Enzyme immobilization by adsorption: A review. *Adsorption, 20*(5, 6), 801–821.

Karboune, S., Neufeld, R., & Kermasha, S., (2005). Immobilization and biocatalysis of chlorophyllase in selected organic solvent systems. *Journal of Biotechnology, 120*(3), 273–283.

Karimpil, J. J., Melo, J. S., & D'Souza, S. F., (2012). Immobilization of lipase on cotton cloth using the layer-by-layer self-assembly technique. *International Journal of Biological Macromolecules, 50*(1), 300–302.

Kawakita, H., Sugita, K., Saito, K., Tamada, M., Sugo, T., & Kawamoto, H., (2002). Production of cycloisomaltooligosaccharides from dextran using enzyme immobilized in multilayers onto porous membranes. *Biotechnology Progress, 18*(3), 465–469.

Kennedy, L. J., Selvi, P. K., Padmanabhan, A., Hema, K. N., & Sekaran, G., (2007). Immobilization of polyphenol oxidase onto mesoporous activated carbons–isotherm and kinetic studies. *Chemosphere, 69*(2), 262–270.

Khan, A. A., & Alzohairy, M. A., (2010). Recent advances and applications of immobilized enzyme technologies: A review. *Research Journal of Biological Sciences, 5*(8), 565–575.

Kirk, O., Borchert, T. V., & Fuglsang, C. C., (2002). Industrial enzyme applications. *Current Opinion in Biotechnology, 13*(4), 345–351.

Krajewska, B., (2009). Ureases. II. Properties and their customizing by enzyme immobilizations: A review. *Journal of Molecular Catalysis B: Enzymatic, 59*(1–3), 22–40.

Krajewska, B., Leszko, M., & Zaborska, W., (1990). Urease immobilized on chitosan membrane: Preparation and properties. *Journal of Chemical Technology and Biotechnology, 48*(3), 337–350.

Krajewska, B., Van, E. R., & Brindell, M., (2012). Temperature-and pressure-dependent stopped-flow kinetic studies of jack bean urease. Implications for the catalytic mechanism. *JBIC Journal of Biological Inorganic Chemistry, 17*(7), 1123–1134.

Kress, J., Zanaletti, R., Amour, A., Ladlow, M., Frey, J. G., & Bradley, M., (2002). Enzyme accessibility and solid supports: Which molecular weight enzymes can be used on solid supports? An investigation using confocal Raman microscopy. *Chemistry–A European Journal, 8*(16), 3769–3772.

Kubáč, D., Čejková, A., Masak, J., Jirků, V., Lemaire, M., Gallienne, E., & Martinkova, L., (2006). Biotransformation of nitriles by *Rhodococcus equi* A4 immobilized in LentiKats®. *Journal of Molecular Catalysis B: Enzymatic, 39*(1–4), 59–61.

Kumar, A., & Kanwar, S. S., (2011). Synthesis of ethyl ferulate in organic medium using celite-immobilized lipase. *Bioresource Technology, 102*(3), 2162–2167.

Leresche, J. E., & Meyer, H. P., (2006). Chemocatalysis and biocatalysis (biotransformation): Some thoughts of a chemist and of a biotechnologist. *Organic Process Research and Development, 10*(3), 572–580.

Liang, J. F., Li, Y. T., & Yang, V. C., (2000). Biomedical application of immobilized enzymes. *Journal of Pharmaceutical Sciences, 89*(8), 979–990.

Mendes, A. A., Oliveira, P. C., Vélez, A. M., Giordano, R. C., De Lc Giordano, R., & De Castro, H. F., (2012). Evaluation of immobilized lipases on poly-hydroxybutyrate beads to catalyze biodiesel synthesis. *International Journal of Biological Macromolecules, 50*(3), 503–511.

Nedovic, V., Leskosek-Cukalovic, I., & Vunjak-Novakovic, G., (1999). *Immobilized Cell Technology (ICT) in Beer Fermentation–a Possibility for Environmentally Sustainable and Cost-Effective Process*. University of Belgrade.

Nguyen, H. H., & Kim, M., (2017). An overview of techniques in enzyme immobilization. *Applied Science and Convergence Technology, 26*(6), 157–163.

Nisha, S., Karthick, S. A., & Gobi, N., (2012). A review of methods, application, and properties of immobilized enzyme. *Chemical Science Review and Letters, 1*(3), 148–155.

Ozturk, T. K., & Kilinc, A., (2010). Immobilization of lipase in organic solvent in the presence of fatty acid additives. *Journal of Molecular Catalysis B: Enzymatic, 67*(3, 4), 214–218.

Park, J. K., & Chang, H. N., (2000). Microencapsulation of microbial cells. *Biotechnology Advances, 18*(4), 303–319.

Pešić, M., López, C., Álvaro, G., & López-Santín, J., (2012). A novel immobilized chloroperoxidase biocatalyst with improved stability for the oxidation of amino alcohols to amino aldehydes. *Journal of Molecular Catalysis B: Enzymatic, 84*, 144–151.

Porath, J., & Axén, R., (1953). In: Mosbach, K., (ed.), *Methods in Enzymology* (Vol. XLIV, pp. 19–45).
Quirk, R. A., Chan, W. C., Davies, M. C., Tendler, S. J., & Shakesheff, K. M., (2001). Poly (l-lysine)–GRGDS as a biomimetic surface modifier for poly (lactic acid). *Biomaterials, 22*(8), 865–872.
Reshmi, R., Sanjay, G., & Sugunan, S., (2006). Enhanced activity and stability of α-amylase immobilized on alumina. *Catalysis Communications, 7*(7), 460–465.
Rinzema, A., Boone, M., Van, K. K., & Lettinga, G., (1994). Bactericidal effect of long chain fatty acids in anaerobic digestion. *Water Environment Research, 66*(1), 40–49.
Rosevear, A., Kennedy, J. F., & Cabral, J. M. S., (1987). Immobilized enzymes and cells. *FEBS Letters, 239*(2), 384, 385.
Sakai, S., Liu, Y., Yamaguchi, T., Watanabe, R., Kawabe, M., & Kawakami, K., (2010). Production of butyl-biodiesel using lipase physically adsorbed onto electrospun polyacrylonitrile fibers. *Bioresource Technology, 101*(19), 7344–7349.
Santos, J. C., Nunes, G. F., Moreira, A. B., Perez, V. H., & De Castro, H. F., (2007). Characterization of *Candida rugosa* lipase immobilized on poly (N-methylolacrylamide) and its application in butyl butyrate synthesis. *Chemical Engineering & Technology: Industrial Chemistry-Plant Equipment-Process Engineering-Biotechnology, 30*(9), 1255–1261.
Sassolas, A., Blum, L. J., & Leca-Bouvier, B. D., (2012). Immobilization strategies to develop enzymatic biosensors. *Biotechnology Advances, 30*(3), 489–511.
Sheldon, R. A., & Van, P. S., (2013). Enzyme immobilization in biocatalysis: Why, what, and how. *Chemical Society Reviews, 42*(15), 6223–6235.
Sheldon, R. A., (2007). Cross-linked enzyme aggregates (CLEAs): Stable and recyclable biocatalysts. *Biochemical Society Transactions, 35*(6), 1583–1587.
Sheldon, R. A., (2007). Enzyme immobilization: The quest for optimum performance. *Advanced Synthesis and Catalysis, 349*(8-9), 1289–1307.
Šulek, F., Fernández, D. P., Knez, Ž., Habulin, M., & Sheldon, R. A., (2011). Immobilization of horseradish peroxidase as crosslinked enzyme aggregates (CLEAs). *Process Biochemistry, 46*(3), 765–769.
Thörn, C., Gustafsson, H., & Olsson, L., (2011). Immobilization of feruloyl esterases in mesoporous materials leads to improved transesterification yield. *Journal of Molecular Catalysis B: Enzymatic, 72*(1, 2), 57–64.
Torres, P., Reyes-Duarte, D., López-Cortés, N., Ferrer, M., Ballesteros, A., & Plou, F. J., (2008). Acetylation of vitamin E by *Candida antarctica* lipase B immobilized on different carriers. *Process Biochemistry, 43*(2), 145–153.
Tufvesson, P., Fu, W., Jensen, J. S., & Woodley, J. M., (2010). Process considerations for the scale-up and implementation of biocatalysis. *Food and Bioproducts Processing, 88*(1), 3–11.
Van, De. V. F., Lourenço, N., Pinheiro, H., & Bakker, M., (2002). Carrageenan: A food-grade and biocompatible support for immobilization techniques. *Advanced Synthesis and Catalysis, 344*, 815–835.
Wong, L. S., Thirlway, J., & Micklefield, J., (2008). Direct site-selective covalent protein immobilization catalyzed by a phosphopantetheinyl transferase. *Journal of the American Chemical Society, 130*(37), 12456–12464.

Wu, C., Zhou, G., Jiang, X., Ma, J., Zhang, H., & Song, H., (2012). Active biocatalysts based on *Candida rugosa* lipase immobilized in vesicular silica. *Process Biochemistry, 47*(6), 953–959.

Xie, T., Wang, A., Huang, L., Li, H., Chen, Z., Wang, Q., & Yin, X., (2009). Recent advances in the support and technology used in enzyme immobilization. *African Journal of Biotechnology, 8*(19).

Xie, W., & Wang, J., (2012). Immobilized lipase on magnetic chitosan microspheres for transesterification of soybean oil. *Biomass and Bioenergy, 36*, 373–380.

Xu, F., Wang, W. H., Tan, Y. J., & Bruening, M. L., (2010). Facile trypsin immobilization in polymeric membranes for rapid, efficient protein digestion. *Analytical Chemistry, 82*(24), 10045–10051.

Yu, J., & Ju, H., (2002). Preparation of porous titania sol− Gel matrix for immobilization of horseradish peroxidase by a vapor deposition method. *Analytical Chemistry, 74*(14), 3579–3583.

Yücel, Y., (2012). Optimization of immobilization conditions of *Thermomyces lanuginosus* lipase on olive pomace powder using response surface methodology. *Biocatalysis and Agricultural Biotechnology, 1*(1), 39–44.

Zaidan, U. H., Rahman, M. B. A., Othman, S. S., Basri, M., Abdulmalek, E., Rahman, R. N. Z. R. A., & Salleh, A. B., (2012). Biocatalytic production of lactose ester catalyzed by mica-based immobilized lipase. *Food Chemistry, 131*(1), 199–205.

Zhai, R., Zhang, B., Liu, L., Xie, Y., Zhang, H., & Liu, J., (2010). Immobilization of enzyme biocatalyst on natural halloysite nanotubes. *Catalysis Communications, 12*(4), 259–263.

Zhang, B., Weng, Y., Xu, H., & Mao, Z., (2012). Enzyme immobilization for biodiesel production. *Applied Microbiology and Biotechnology, 93*(1), 61–70.

CHAPTER 11

Waste Treatment

Waste is an inevitable by-product of the majority of human action. Economic growth and increasing living standards have resulted in a rise in the quantity and complexity of waste generated. In contrast, industrial diversification and increased healthcare services have introduced significant amounts of industrial hazardous waste and biomedical waste into the waste stream, potentially posing severe environmental and human health risks. Waste is generated in various shapes and sizes, and its characteristics can be represented in multiple ways. The physical state of waste, source of generation, recyclable possibilities, biodegradable abilities, and environmental impact are typical characteristics used in waste classification. According to White et al. (1995), based on the physical state of waste, it can be divided into three categories, i.e., solid, liquid, and gaseous wastes. Although different nations have different classifications, the most frequently used categorizations are demonstrated here:

- Based on the physical state of the waste:
  - Solid waste;
  - Liquid waste;
  - Gaseous waste.
- Based on the source of waste generation:
  - Domestic waste;
  - Industrial waste;
  - Agricultural waste;
  - Institutional/commercial waste;
  - Demolition and construction waste;
  - Processing waste.
- Based on the impact of environment:
  - Hazardous waste;
  - Non-hazardous waste.

Handbook of Industrial Food Microbiology. Manju Nehra & Vikash Nain
© 2025 Apple Academic Press, Inc. Co-published with CRC Press (Taylor & Francis)

11.1 SOLID WASTE

Solid waste is described by Tchobanoglous et al. (1993) as waste created by human activities that are solid or semisolid and discarded as useless materials. According to Beranek (1992), solid wastes are a general category of wastes generated from different practices such as agricultural, landscaping, and other processes such as domestic and commercial. According to him, solid pollutants are distinct from wastewater and gaseous waste pollution. Solid wastes, in his opinion, are any wastes that people might generally deem suitable for disposal on the soil. According to Basu (2010), municipal waste, biomedical waste, electronic waste, and toxic waste are all included in solid waste. According to Tchobanoglous et al. (1993), unlike other wastes, solid wastes do not disappear quickly and persist for long periods and are found in the same place where they are discarded. Few common sources of solid waste are municipal solid waste, agricultural solid waste, industrial waste, construction waste and industrial waste. These include electronic products, batteries, tires, paper, cardboard, metal, plastic, food waste, wood, glass, etc. (Table 11.1).

TABLE 11.1 Major Sources and Types of Solid Wastes Generated

Major Source	Typical Solid Wastes
Residential	Food wastes, paper, cardboard, plastics, textiles, leather, yard wastes, wood, glass, metals, ashes, special wastes (e.g., bulky items, consumer electronics, white goods, batteries, oil, tires), and household hazardous wastes.
Construction and demolition	Wood, steel, concrete, dirt, etc.
Municipal	Street sweepings, landscape, and tree trimmings, general wastes from parks, beaches, and other recreational area, sludge.
Industrial	Housekeeping wastes, packaging, food wastes, construction, and demolition materials, hazardous wastes, ashes, special wastes.
Commercial/ Institutional	Paper, cardboard, plastics, wood, food wastes, glass, metals, special wastes, hazardous wastes, etc.
Processing sector	Industrial process wastes, scrap materials, off-specification products, slag, tailings, etc.
Agriculture	Spoiled food wastes, agricultural wastes, hazardous wastes (e.g., pesticides).
Biomedical waste	Cotton, needles, syringes, discarded medicines, dressing, blades, glass, scalpels, etc.

11.2 SOLID WASTE MANAGEMENT

11.2.1 LANDFILLING

As industrial waste is transferred off-site, it is usually deposited in a landfill. Daily, the waste is tipped into a massive landfill (pit or trench) in the land and is then covered with excavated soil. To prohibit animals from scratching up the waste and insects' growth, approximately 0.5 m of soil layer is used to cover the accumulated waste every day. The site of landfills should be determined in conjunction with local councils and the people that would be impacted. Ideally, places should be walled and located at least one kilometer downstream of nearby communities. The requirement of a large area and its effect on the nearby population are associated drawbacks of solid waste management by landfilling technique.

11.2.2 INCINERATION

While burning or incineration is commonly used to dispose of combustible waste, it should be done off-site or at least a long distance away from residential area. Burning garbage in residential areas may pose a significant smoke or fire threat, mainly if several fires are ignited simultaneously. Incineration may minimize waste volume, which can be suitable for burial or landfill space. Waste can be incinerated in pits and should be covered with good sail as practiced in landfilling technique. Like landfill method, this technique too has the limits of availability of adequate area for incineration site. Burning significantly reduces the quantity of waste materials, and it is often appropriate for reducing scavenging in off-site dumps. There could be a risk of smoke or burning by this technique and it a significant limitation of this method (Figure 11.1).

FIGURE 11.1 Different methods of solid waste management.

11.2.3 COMPOSTING

Composting vegetables and other organic waste are a simple process used in various circumstances. Food waste may be dug into the field to incorporate humus and fiber where people have their gardens or flower plots. This ensures that the waste is entirely safe; also, the fertility of the soil is benefited. This method should be preferred wherever circumstances favor. Sufficient moisture and chemical level selecting microbial activity is the prerequisite of this process. A close monitoring and regulation of these level speeds up the composting procedure. This is meant to promote compost that is easy to handle and efficient as a fertilizer. Such programs require a significant amount of knowledge and skill and are ideally operated from a centralized spot. Composting is beneficial for both the climate and agriculture. Associated limitation of this method is that large-scale projects necessarily require intensive supervision and skilled staff.

11.2.4 RECYCLING

Complex recycling schemes are unlikely to be appropriate, although recycling certain waste materials is necessary on certain possible occasions. Various waste materials such as plastic bags, glass, and tin vessels are generated on specific circumstances; they must be recycled for efficient management. Recycling reduces the load of waste management and provides raw material for reuse of that particular commodity. This technique must be adopted for well-being of human and to minimize the impact of waste on the ecosystem. This technique is most eco-friendly, but expensive processing restricts its application to certain limits.

11.3 WASTEWATER

Access to safe drinking water is regarded as a fundamental human right denied to large sections of the world's population. Its accessibility has a significant effect on human existence. As a consequence of economic and population growth, water scarcity is regarded as one of humanity's greatest fears and a stumbling block to long-term sustainability. There are just a few strategies for dealing with the problems of freshwater scarcity; these options include distillation, water conservation projects, recycling, and reuse of wastewater and conservation of rainfall water. The amount and quality of waste produced and discharged into natural water bodies has dominated headlines in various international media outlets as a major cause of concern.

The wealthiest countries are devoting significant resources to reducing pollution caused by large populations. Wastewater is primarily made up of unwanted waste that could pollute our water system, such as sewage sludge, wastewater from domestic, industrial, dairy, and fish canning operations, and waste from the oil and gas industry. Water contamination is caused by wastewater from manufacturing and chemical refining sectors. Specific chemical compounds that are easily traceable are usually found in industrial wastewater. As a result, several wastewater treatment approaches have been devised and continually developed over time as research laboratories worldwide have improved. Wastewater treatment technologies can be categorized into biological, chemical, and physical treatment methods, and further classified as in-situ or ex-situ technologies, depending on the classifications of the concerned wastewater source (Figure 11.2).

FIGURE 11.2 Major sources of water pollution.

11.3.1 MAJOR WATER CONTAMINANTS

Water is contaminated and polluted as a result of domestic and industrial use. Contaminants are chemicals that may cause diseases and render water unfit for human use. In contrast, contaminants reduce the utility of water or render it unpleasant to the sensory organs, taste, and smell. Pollution can be accompanied by contamination. Contaminants may be included in domestic wastewater, and agricultural waste can produce contaminants and pollutants. Major water contaminates/pollutants are listed here:

1. **Heavy Metals:** These are a source of contaminants of groundwater. A lab analysis of the water is the safest way to detect their presence. High amounts of heavy metals in drinking water have raised questions over chronic toxicity.
2. **Turbidity:** It relates to the amount of suspended solids in a liquid; for example, muddy water is exceptionally turbid. Turbidity is undesirable because it reduces the efficiency of water treatment approaches by shielding pathogens from chemical or thermal destruction (TD), or by absorbing UV light itself in the case of UV treatment.
3. **Inorganic Pollutants:** Mineral acids, alkalis, inorganic salts, hydrogen sulfide, ammonia, free chlorine, salts of chromium, nickel, brass, cadmium, copper, silver, and anions such as sulfates, phosphates, chlorides, cyanides, nitrites, and nitrates, and cations such as calcium, sodium, magnesium, iron, potassium, mercury, arsenic, manganese are commonly found in municipal and industrial waste water.
4. **Organic Compounds:** Various organic substances such as chloroform, oil, poisons, and herbicides can contaminate water. Other organic contaminants include high molecular weight compounds includes sugar, oils, fats, protein, hydrocarbons, phenols, detergents, and organic acid from industrial and agricultural processes. Few of these contaminants are biodegradable, while others are highly toxic to aquatic organisms. To treat such wastewater through biological methods, removing or reducing such substances to a particular concentration is necessary. Food and agro-product industries, such as distilleries, sugar factories, dairies, fruit canning, meat processing, and pulp and paper mills, account for 65–70% of the organic contaminants released in water bodies in India.
5. **Pathogens:** Protozoa, bacteria, and viruses are the major pathogens that are predominantly found in wastewater. Protozoa are the most

common contaminants in drinking water, and they cause several diseases. Bacteria are more diminutive than protozoa and cause various diseases, namely typhoid fever, cholera, diarrhea, and dysentery. Viruses are the second most dangerous pathogen after protozoa. Many waterborne infectious caused by virus do not show a lethal threat to a healthy human.

11.4 BIOLOGICAL WASTE TREATMENT

The type of microorganisms and their quantity may vary based on type of wastewater treated. Most industrial operations generate waste waters with different concentrations of salts and organic compounds. The manufacturing industries in general release substances which are biodegradable. Often, these wastes can be utilized and recycled by microorganisms and plants. Agriculture and Dairy industries produce crop residues and manures, which are biodegradable. Water soluble, colloidal, and suspended pollutants are found in the waste waters generated from industrial fermentation processes.

Generally, the bacteria present in the sewage include coliforms, fecal streptococci, anaerobic spore-forming bacilli, proteus group, etc. Depending on its size and the type of waste waters it produces, a fermentation company may process its waste waters or arrange with the local municipal sewerage for treatment at the local sewage treatment plant. The natural microbial floras of the waste waters include heterotrophic bacteria, to certain extent, actinomycetes, and fungi. Protozoa also are very active in organic matter decomposition. Anaerobic bacteria are helpful in the anaerobic digestion of organic matter.

Sewage treatment on a large scale of population of towns and cities is known as large scale treatment. This sewage is treated by municipal plants. The overall treatment process is categorized in three steps:

- Primary treatment;
- Secondary treatment; and
- Tertiary treatment.

11.4.1 PRIMARY TREATMENT

This step aims on removal the settle able solids present in the wastewater. The solids are removed by screening or grinding. Sludge recovered from this step is treated with any suitable method described in the solid waste management section. Sedimentation tanks are used for primary treatment and are subjected to aluminum sulfate to entrap the organic compounds and microorganisms. The volume, shape, and specific gravity of the particles are the major factors that significantly affect sedimentation process. The sedimentation tanks are constructed based on sewage flow, shape, purpose, position, and operation. Around 60–80% of the suspended particles can be separated using chemical precipitation. Alum (aluminum sulfate), iron salts (ferric sulfate, ferrous sulfate, ferric chloride), lime and soda ash (sodium carbonate), sodium silicate, and sodium aluminate are the most widely used coagulants in water treatment.

11.4.2 SECONDARY TREATMENT (BIOLOGICAL)

It is carried to absorb and oxidize organic constituents of wastewater, thus reducing BOD. The procedure involves the oxidation of organic material in the wastewater by microbial activity. Microorganisms such as bacteria, algae, fungi, protozoa, rotifers, and nematodes are used in this phase to decompose the unstable organic matter into a stable inorganic phase. The microbial activity may be aerobic or anaerobic.

The methods employed are:

- Filtration by trickling filters;
- Activated sludge process;
- Use of oxidation ponds;
- Anaerobic digesters.

11.4.2.1 TRICKING FILTERS

A tricking filter comprises a large bed of stone pieces, slag or synthetic material with drains made at the bottom of the tank. The sewage is made to slowly trickle over this pile of stones by a rotating arm or sprinklers. Due to the spraying process, sewage is saturated by oxygen. Due to aerobic condition, microbial flora will develop in the bed (bacteria, fungi, protozoa, and algae). This microbial growth is generally referred to as Zoogloea film or biological film, and as the sewage trickles through it, microorganisms metabolize organic constituents. This film has a thickness of 0.1 to 2 mm and is rich in microorganisms. Few weeks are required for the microbial film to develop in the trickling filter. The biological film is rich in the bacteria namely Beggiatoa alba, Sphaerotilus natons, Achromobacter spp., Pseudomonas, and algae namely chlorella, Utotrix and Stigeochlonium and some fungi and yeast. The upper region of the tricking filter is favorable for growth of algae. If too many algae are allowed to grow, other microorganisms may be suppressed, and efficiency of the filter may be reduced (Figure 11.3).

Tricking Filters **Activated Sludge Process**

Activate

FIGURE 11.3 Tricking filters and activated sludge process.

Source: Modified from Netsol Water Solutions. https://www.netsolwater.com/an-insight-on-activated-sludge-process,-loading-rate.php?blog=2275

11.4.2.2 ACTIVATED SLUDGE PROCESS

In this method, the wastewater is subjected to vigorous aeration, resulting in flock formation. The sewage is passed onto an aeration tank from primary settling tank. Sewage is aerated by mechanical stirring. The fine colloidal particles along with microorganisms form aggregates known as flocculus. The flocks are permitted to settle down secondary settling tank. Activated sludge contains actively metabolizing bacteria, yeast, molds, and protozoa. The activated sludge process reduces the BOD of effluent to 10–15% compared to raw sewage. The advantages of activated sludge process are significant reduction in BOD and suspended solids, low cost, requirement of little land, reduction in intestinal pathogens. Some of the bacteria present in activated sludge flocks are Pseudomonas, E. coli, Achromobacter, Enterobacter, Zooglea, Arthrobacter, Mycobaterium, Corynebacterium, and large filamentous bacteria, fungi, and protozoa.

11.4.2.3 OXIDATION PONDS

Oxidation ponds are used for secondary treatment in rural areas or industrial sectors. The oxidation ponds (aerated ponds, stabilization ponds) are shallow 2–4 feet in depth, designed to allow algal growth. The organic materials are degraded by heterotrophic bacteria into simpler forms; they support algae growth. Chlorella pyrenoidosa, C. ellipsoides, Scendesmus acutus, S. quadricauda are algae growing in stabilization ponds.

11.4.2.4 ANAEROBIC DIGESTERS

The sludge which accumulates in aerobic sewage treatment is treated in anaerobic digesters through the process of anaerobic digestion. Anaerobic digestion is primarily helpful for the stabilization of concentrated sludge, and the process is carried out in an airtight reactor. The microorganisms degrade sludge into soluble substances and gaseous products. Sludge is introduced into the digester continuously or intermittently. These digesters are provided with provisions for mechanical mixing, heating, gas collection, sludge addition and removal of final stabilized sludge.

Anaerobic digestion involves the following steps:

1. **Fermentation:** The formation of organic acids from organic polymers is due to the bacteria fermentation of sludge components (Clostridum, Eubacterium, Lactobacillus, Peptostreptococus, etc.) in the digester. The organic acids produced are butyrate, propionate, succinate, lactate, acetate, ethanol, and hydrogen, CO2, etc.
2. **Acidogenesis:** The organic acids produced during fermentation are converted to acidic products (acetate) by several acidogenic bacteria viz., Syntrophomonas, Syntrophobacter and Acidobacterium.
3. **Methanogenesis:** This is the last stage and involves utilizing products produced during acidogenesis as substrates by methanogenic bacteria and produces methane and carbon dioxide (CO2). Methane gas is highly insoluble and is released from the digester represents the stabilization of sewage or sludge. The products obtained after anaerobic digestion are a mixture of gases (70% CH4, 30% CO2), microbial biomass and non-biodegradable compounds like heavy metals, polychlorinated biphenyl, etc.

11.4.3 TERTIARY TREATMENT

It is needed to remove non-biodegradable organic material, heavy metals and minerals. Tertiary treatment involves removing solids, biological nitrogen, biologic phosphorous, and killing of microorganisms (disinfection). The solids can be removed by granular medium filtration and microscreening methods. The nitrogen is released in nitrate by converting to nitrogen gas (denitrification). Acinetobacter species are employed to remove phosphorous by phosphate stripper and process system. Disinfection kills the microorganisms, and this process is necessary from the point of view of public health as most of the effluents are discharged to rivers and other water bodies. Physical agents like heat and light are used as disinfectants, and the chemicals like bromine, ozone, iodine, phenols, alcohols, alkalis, acids are also used as disinfectants. If disinfection is not done, the effluents may affect water supply systems downstream by harming irrigation and fish harvesting. Chlorination (chlorine) is the usually employed method of disinfection.

11.4.4 MICROBIOLOGY OF BIOLOGICAL WASTEWATER TREATMENT

Biological wastewater treatment technology is the most effective treatment method. Biodegradation is a term that may describe biological wastewater treatment technology. Biodegradation is the decay of organic materials by microorganisms such as bacteria, fungi, and microflora, leading to CO2, water, and methane. Microorganisms are undoubtedly the most significant factor in pollutant degradation and maintaining biological system stability. Biological wastewater treatment systems have much strength over other treatment methods, including lower costs, minimum or no secondary pollution excretion, and, most significantly, fewer negative environmental impacts. Furthermore, biological wastewater treatment technology has the advantage over chemical and other traditional wastewater treatment methods in terms of operating costs and capital expenditures.

11.4.5 MICROORGANISMS ASSOCIATED IN WASTEWATER TREATMENT

Microalgae can transform solar energy into biomass and have shown the potential to consume nutrients such as nitrogen and phosphorus in a limited period. Ankistrodesmus, Golenkinia, Euglena, and Micractinium are a few of the genera of algae associated with pond stabilization. These microalgae may be used to manage a wide variety of sewage systems, and they are considered a tertiary treatment method for removing organic ions. Organic ion removal can be accomplished chemically or naturally, with biomedical therapy seeming to be more promising. In wastewater treatment systems, heterotrophic bacteria eliminate the organic matter. These bacteria function in clusters as floc, biofilm, or granules during wastewater treatment.

Protozoa play a significant role in removing organic matter from sewage, but their primary predatory activity is bacteria in liquid media. Apart from microbes, protozoa are involved in activating sludge in wastewater treatment. Protozoa also help to sustain biological ecosystem equilibrium by removing excess bacteria, promoting their growth, and promoting floc. As bacteria accelerate the carbon supply, protozoa may mineralize carbon in activated sludge and remove mineral nutrients. On the other hand, Protozoa can secrete growth-stimulating compounds that encourage bacterial activity.

11.5 ENZYME IN WASTE TREATMENT

The majority of waste treatment processes may be classified as a physicochemical or biological processes. Since it includes chemical processes dependent on the activity of biological catalysts, enzymatic treatment falls within these two traditional categories. The following are some of the potential benefits of enzymatic treatment over conventional treatment: application of bio-refractory compounds; capability to operate at both high and low concentration of contaminants; operation over a wide range of pH, temperature, and salinity; absence of shock loading effects; absence of delays associated with biomass acclimation; reduction in sludge volume (no biomass generation). Food manufacturing is a huge sector that generates a significant amount of waste. On the other hand, food wastes benefit from being susceptible to transformation into food, feed, or non-food items with added value.

In contrast, wastes produced by other factories are usually harmful and must be made harmless by adequate treatment schemes. The use of enzymes in treating food waste has recently been proposed. Enzymes may reduce food waste by converting its enzymatic activity to provide higher-value added-by-products and further clean up food waste channels.

Proteases are hydrolases often used to treat fish and meat waste in the food sector. Proteins in waste streams may be solubilized by proteases, resulting in recoverable liquid concentrates or dry solids of nutritional value for fish or livestock. Proteases break down insoluble proteins through a multistep mechanism in which the enzyme, which is initially adsorbed on the solid substrate, cleaves strongly attached polypeptide chains. Then, the rate of solubilization will be slower if the substrate's core is more compact, depending on the dilution of the enzyme to surface active sites and core particles. Dalev (1994) reported using alkaline proteinase from Bacillus subtilis to process waste feathers from poultry slaughterhouses. Feathers make up 5% of the body weight of poultry and may be used as a high-protein food and feed supply if the rigid keratin matrix is entirely broken. After pretreatment with NaOH, mechanical disintegration, the feathers were fully solubilized and enzyme hydrolysis. The finished product was a dense, grayish powder with a high protein content that could primarily be used as a feed ingredient.

Amylases are polysaccharide hydrolases used to manage starch-containing food wastewaters and saccharify and ferment starch. Amylases, according to Shoemaker, may be used to make alcohol from rice processing wastewater. Amylases also improved activated sludge wastewater treatment by shortening the treatment time. Coleman described the unique use of α-amylase and glucoamylase in manufacturing photodegradable and biodegradable plastics. The method involves converting starchy materials like cheese whey or potato waste from industrial food production into fully biodegradable plastics. The long starch molecule is broken down into smaller pieces by amylase. The latter are then targeted by glucoamylase, which causes saccharification and glucose production (more than 90% of the starch is converted to glucose). A strain of lactic acid bacteria (LAB) ferments the glucose into lactic acid (LA). Finally, LA is recovered, refined, and used to take photo- and biodegradable plastics, the rate of which can be regulated by a judicious combination of different LA isomers and other compounds. The final product contains 95% LA and 5% of an environmentally friendly product.

Aromatic compounds, such as phenols and aromatic amines, are among the most common contaminants and are severely regulated in many countries. They can be present in wastewater from many sectors, including coal conversion, petroleum processing, resins, and plastics. Here, we can use the enzymes that can successfully degrade the phenolic compounds. A variety of microorganisms and plants contain peroxidases, which are oxidoreductase enzymes. They are all activated by the presence of peroxides, such as hydrogen peroxide. Once activated by H2O2, Horseradish peroxidases can catalyze the oxidation of a wide variety of toxic aromatic compounds such as phenols, biophenols, anilines, benzidines, and related hetero aerometric compounds. In addition, horseradish peroxidase can co-precipitate such pollutants that are impossible to isolate. Plant peroxidase extracted from tomato and water hyacinth plants, Chloroperoxidase from the fungus Caldariomyces fumago, Manganese peroxidase developed by Phanerochete chrysosporium, Microbial peroxidase from Coprinus macrorhizus can act as an alternative to horseradish peroxidases.

Another category of oxidoreductase that has been found to catalyze the oxidation of phenolic compounds is polyphenol oxidases. They are classified into two categories: Laccase is produced by a variety of fungi and tends to have the ability to reduce the toxicity of phenolic compounds through a polymerization process. Tyrosinase, also called polyphenol oxidase, phenolase, or catecholase, catalyzes two reactions in series:

- The formation of diphenols by hydroxylation of monophenols with molecular oxygen;
- The formation of o-quinones by dehydrogenating o-diphenols with oxygen.

In the manufacturing of chemical intermediates, synthetic fibers, rubber, pharmaceuticals, ore leaching, coal processing, and metal plating, it is estimated that about 3 million tons of cyanide are used annually throughout the world. Since cyanide is a metabolic inhibitor that can damage humans and other animals, it must be eliminated from effluents before discharge. Since cyanide is a metabolic inhibitor that can damage humans and other animals, it must be eliminated from effluents before discharge. Cyanidase is a type of enzyme that breaks down. Cyanide is a novel enzyme development that intends to convert cyanide in industrial wastewaters to ammonia in a single-step reaction. Cyanidase is obtained from gram-positive bacteria strains isolated from the Alcaligenes denitrificans genus and is produced using proprietary methods. It has been discovered that cyanide hydratase, also known as formamide hydrolase, hydrolyzes cyanide into formamide. Several fungi develop cyanide hydratase, which can be induced by pre-exposing the fungi to low levels of cyanide.

Lipases enzyme obtained from microbial sources showed a capability of catalyzing various reactions such as hydrolysis, inter-esterification, esterification, alcohol-lysis, and amino-lysis. Lipase has many possible uses in the dairy, pharmaceutical, detergent processing, cosmetic, and paper producing sectors, in addition to its diagnostic use in bioremediation, but its high production cost has limited its industrial use (Sharma et al., 2011). Cellulases break down cellulose into reducing sugars that yeasts or bacteria can further transform into ethanol during enzymatic hydrolysis in the fermentation process. Cellulases are used in the paper and pulp industries to remove ink from recycled paper.

11.5.1 ADVANTAGES OF ENZYMATIC TREATMENT OVER OTHER TECHNIQUES

The constantly intensifying application of enzymes generates a mounting demand for novel biocatalysts with improved or innovative properties. Enzymes can eliminate truculent toxins by precipitation or transformation into other materials by acting directly on them. Biological (enzymatic) systems benefit conventional chemical/physical processes in that they are considered safe and environmentally friendly. Chemical precipitation, coagulation, flocculation, floatation, and membrane filtration are examples of physicochemical treatments with advantages such as ease of operation and control, temperature flexibility, and efficiency. Still, their advantages are counterbalanced by many disadvantages such as high operational costs due to the chemicals used, high energy consumption, and handling cost. Enzymes are catalysts that are highly specific and effective. They can selectively degrade a pollutant without disturbing the effluent's other constituents. They can work under various reaction conditions, including temperature and pH. High costs of sophisticated chemical treatments for wastewater treatment, production of additional solid waste, and strict government policies prompted the development of many successful and easy biological methods for waste degradation.

Enzymes are more environmentally friendly owing to their biodegradability nature. The enzyme accepts one or two electrons from the substrate and donates them to an electron acceptor in reactions where the target pollutant is oxidized. As a result, the enzyme is recovered after the reaction and available for the next catalytic step. Enzymes' biological origin eliminates their adverse environmental impact, rendering enzymatic wastewater treatment an environmentally friendly process.

KEYWORDS

- **activated sludge process**
- **anaerobic digesters**
- **biological waste treatment**
- **enzymatic treatment**
- **landfilling**
- **oxidation ponds**
- **recycling**
- **solid waste management**
- **tricking filters**

REFERENCES

Abdel-Raouf, N., Al-Homaidan, A. A., & Ibraheem, I. B. M., (2012). Microalgae and wastewater treatment. Saudi Journal of Biological Sciences, 19(3), 257–275.

Adam, W., Lazarus, M., Saha-Möller, C. R., Weichold, O., Hoch, U., Häring, D., & Schreier, P., (1999). Biotransformations with peroxidases. Biotransformations, 73–108.

Anjaneyulu, Y., Chary, N. S., & Raj, D. S. S., (2005). Decolourization of industrial effluents–available methods and emerging technologies: A review. Reviews in Environmental Science and Bio/Technology, 4(4), 245–273.

Basu, R., (2009). Solid waste management-a model study. Sies Journal of Management, 6, 20–24.

Beranek, W., (1992). Solid waste management and economic development. Economic Development Review, 10, 49.

Blaschek, H. P., (1992). Approaches to making the food processing industry more environmentally friendly. Trends in Food Science & Technology, 3, 107–110.

Coleman, R., (1990). Biodegradable plastics from potato waste double savings to environment. Agricultural Engineering; (USA), 71(6).

Crini, G., & Badot, P. M., (2007). Traitement et Épuration des Eaux Industrielles Polluées (pp. 353). PUFC, Besançon.

Crini, G., & Badot, P. M., (2010). Sorption Processes and Pollution: Conventional and Non-Conventional Sorbents for Pollutant Removal from Wastewaters. Presses Univ. Franche-Comté.

Dadrasnia, A., Usman, M. M., Lim, K. T., Velappan, R. D., Shahsavari, N., Vejan, P., & Ismail, S., (2017). Microbial aspects in wastewater treatment–a technical. Environmental Pollution and Protection, 2(2), 75–84.

Dalev, P. G., (1994). Utilization of waste feathers from poultry slaughter for production of a protein concentrate. Bioresource Technology, 48(3), 265–267.

De Carvalho, C. C., (2011). Enzymatic and whole cell catalysis: Finding new strategies for old processes. Biotechnology Advances, 29(1), 75–83.

Demirbas, A., (2011). Waste management, waste resource facilities and waste conversion processes. Energy Conversion & Management, 52(2), 1280–1287.

Dixon, N., & Jones, D. R. V., (2005). Engineering properties of municipal solid waste. Geotextiles & Geomembranes, 23(3), 205–233.

Ferrera, I., & Sanchez, O., (2016). Insights into microbial diversity in wastewater treatment systems: How far have we come? Biotechnology Advances, 34(5), 790–802.

Goplerud, III. C. P., (1995). Water pollution law: Milestones from the past and anticipation of the future. Natural Resources & Environment, 10, 7–12.

Grady, Jr. C. L., Daigger, G. T., Love, N. G., & Filipe, C. D., (2011). Biological Wastewater Treatment. CRC press.

Gray, N. F., (2005). Water Technology: An Introduction for Environmental Scientists and Engineers (2nd edn.). Elsevier Science & Technology Books, ISBN 0750666331, Amsterdam, The Netherlands.

JA, N., Al-Kassim, L., Bewtra, J. K., & Taylor, K. E., (1993). Wastewater treatment by enzyme-catalyzed polymerization and precipitation. In: Biodeterioration Abstracts (Vol. 7, No. 1, pp. 1–8).

James, M., (1985). Montgomery Consulting Engineers, Inc. Water Treatment Principles and Design.

Kaila, S., Sejpal, P., & Mona, M. B., (2018). Application of various enzymes in waste treatment. International Journal of Scientific Research in Chemistry, 3(1), 60–64.

Karam, J., & Nicell, J. A., (1997). Potential applications of enzymes in waste treatment. Journal of Chemical Technology & Biotechnology: International Research in Process, Environmental & Clean Technology, 69(2), 141–153.

Klibanov, A. M., (1982). Enzymatic removal of hazardous pollutants from industrial aqueous effluents. In: Enzyme Engineering (pp. 319–324). Springer, Boston, MA.

Kurniawan, T. A., Chan, G. Y., Lo, W. H., & Babel, S., (2006). Physico–chemical treatment techniques for wastewater laden with heavy metals. Chemical Engineering Journal, 118(1, 2), 83–98.

Lokhande, R. S., Singare, P. U., & Pimple, D. S., (2011). Toxicity study of heavy metals pollutants in wastewater effluent samples collected from Taloja industrial estate of Mumbai, India. Resources and Environment, 1(1), 13–19.

Madoni, P., (2011). Protozoa in wastewater treatment processes: A minireview. Italian Journal of Zoology, 78(1), 3–11.

Mingjun, S. H. A. N., Yanqiu, W. A. N. G., & Xue, S. H. E. N., (2009). Study on bioremediation of eutrophic lake. Journal of Environmental Sciences, 21, S16–S18.

Mutambanengwe, C., Oyekola, O., & Togo, C., (2008). Production of Enzymes for Industrial Wastewater Treatment–Proof of Concept and Application to the Textile Dye Industry. http://hdl.handle.net/11189/5367.

Nasr, M., Elreedy, A., Abdel-Kader, A., Elbarki, W., & Moustafa, M., (2014). Environmental consideration of dairy wastewater treatment using hybrid sequencing batch reactor. Sustainable Environment Research, 24(6), 449–456.

Nelson, D. L., & Cox, M. M., (2005). Lehninger. Principles of Biochemistry (pp. 47–50). W.H. Freeman, New York.

Ng, K. K., Shi, X., Ong, S. L., & Ng, H. Y., (2016). Pyrosequencing reveals microbial community profile in anaerobic bio-entrapped membrane reactor for pharmaceutical wastewater treatment. Bioresource Technology, 200, 1076–1079.

Nicell, J. A., (1994). Kinetics of horseradish peroxidase-catalyzed polymerization and precipitation of aqueous 4-chlorophenol. Journal of Chemical Technology & Biotechnology: International Research in Process, Environmental AND Clean Technology, 60(2), 203–215.

Nicell, J. A., Bewtra, J. K., Biswas, N., St. Pierre, C. C., & Taylor, K. E., (1993). Enzyme catalyzed polymerization and precipitation of aromatic compounds from aqueous solution. Canadian Journal of Civil Engineering, 20(5), 725–735.

Nicell, J. A., Bewtra, J. K., Taylor, K. E., Biswas, N., & St. Pierre, C., (1992). Enzyme catalyzed polymerization and precipitation of aromatic compounds from wastewater. Water Science and Technology, 25(3), 157–164.

Oh, S. T., Kim, J. R., Premier, G. C., Lee, T. H., Kim, C., & Sloan, W. T., (2010). Sustainable wastewater treatment: How might microbial fuel cells contribute. Biotechnology Advances, 28(6), 871–881.

Pandey, K., Singh, B., Pandey, A. K., Badruddin, I. J., Pandey, S., Mishra, V. K., & Jain, P. A., (2017). Application of microbial enzymes in industrial wastewater treatment. International Journal of Current Microbiology and Applied Sciences, 6, 1243–1254.

Prasad, M. P., & Manjunath, K., (2011). Comparative Study on Biodegradation of Lipid-Rich Wastewater Using Lipase Producing Bacterial Species, 10(2), 121–124.

Rajagopal, R., (2008). Treatment of Agro-Food Industrial Wastewaters Using UAF and Hybrid UASB-UAF Reactors. Doctoral dissertation, Ph. D. Thesis, Indian Institute of Technology Roorkee, Roorkee, India.

Rajasulochana, P., & Preethy, V., (2016). Comparison on efficiency of various techniques in treatment of waste and sewage water–a comprehensive review. Resource-Efficient Technologies, 2(4), 175–184.

Russell, D. L., (2006). Practical Wastewater Treatment. John Wiley & Sons, Inc., ISBN-13: 978-0-471-78044-1, Hoboken, New Jersey, USA.

Samer, M., (2015). Biological and chemical wastewater treatment processes. Wastewater Treatment Engineering, 150.

Samer, M. (Ed.). (2015). Wastewater Treatment Engineering. InTech. doi: 10.5772/59384.

Seghezzo, L., Zeeman, G., Van, L. J. B., Hamelers, H. V. M., & Lettinga, G., (1998). A review: The anaerobic treatment of sewage in UASB and EGSB reactors. Bioresource Technology, 65(3), 175–190.

Sharma, D., Sharma, B., & Shukla, A. K., (2011). Biotechnological approach of microbial lipase: A review. Biotechnology, 10(1), 23–40.

Shoemaker, S., (1986). The Use of Enzymes for Waste Management in the Food Industry (pp. 259–267). Nayes Publications, Park Ridge, NJ.

Singare, P. U., Lokhande, R. S., & Naik, K. U., (2010). A case study of some lakes located at and around thane city of Maharashtra, India, with special reference to physico-chemical properties and heavy metal content of lake water. Interdisciplinary Environmental Review, 11(1), 90–107.

Sponza, D. T., & Uluköy, A., (2005). Treatment of 2, 4-dichlorophenol (DCP) in a sequential anaerobic (upflow anaerobic sludge blanket) aerobic (completely stirred tank) reactor system. Process Biochemistry, 40(11), 3419–3428.

Sun, Y., & Cheng, J., (2002). Hydrolysis of lignocellulosic materials for ethanol production: A review. Bioresource Technology, 83(1), 1–11.

Tchobanoglous, G., Theisen, H., & Vigil, S., (1993). Integrated solid waste management: Engineering principles and management issues. Water Science & Technology Library, 8(1), 63–90.

Venugopal, V., Alur, M. D., & Nerkar, D. P., (1989). Solubilization of fish proteins using immobilized microbial cells. Biotechnology and Bioengineering, 33(9), 1098–1103.

White, G. C., (1972). Handbook of Chlorination. Van Nostrand Reinhold, New York.

White, P. R., Franke, M., & Hindle, P., (1995). Integrated Solid Waste Management: A Lifecycle Inventory. Berlin: Springer.

Index

B

D

G

J

K

Q

R

S

W

X

Y

Z